Motivation of New Generation Students for Learning Physics and Mathematics

BALTISCHE STUDIEN ZUR ERZIEHUNGS- UND SOZIALWISSENSCHAFT

Herausgegeben von Gerd-Bodo von Carlsburg,
Algirdas Gaižutis und Airi Liimets

BAND 33

Zu Qualitätssicherung und Peer Review der vorliegenden Publikation

Die Qualität der in dieser Reihe erscheinenden Arbeiten wird vor der Publikation durch die Herausgeber der Reihe geprüft.

Notes on the quality assurance and peer review of this publication

Prior to publication, the quality of the work published in this series is reviewed by editors of the series.

Palmira Pečiuliauskienė / Valdemaras Aleksa

Motivation of New Generation Students for Learning Physics and Mathematics

PETER LANG

Bibliografische Information der Deutschen Nationalbibliothek
Die Deutsche Nationalbibliothek verzeichnet diese Publikation in der Deutschen Nationalbibliografie; detaillierte bibliografische Daten sind im Internet über http://dnb.d-nb.de abrufbar.

Library of Congress Cataloging-in-Publication Data
A CIP catalog record for this book has been applied for at the Library of Congress

Reviewers:
Andris Broks, Doctor of Physics,
Professor emeritus of the University of Latvia
Konstantinos D. Chatzidimou, Dr., Assist. Professor,
Aristotle University of Thessaloniki

ISSN 1434-8748
ISBN 978-3-631-74611-0 (Print)
E-ISBN 978-3-631-76332-2 (E-PDF)
E-ISBN 978-3-631-76333-9 (EPUB)
E-ISBN 978-3-631-76334-6 (MOBI)
DOI 10.3726/b14474

Peter Lang – Berlin · Bern · Bruxelles · New York ·
Oxford · Warszawa · Wien

Diese Publikation wurde begutachtet.

www.peterlang.com

Contents

0 Introduction

By Palmira Pečiuliauskienė

The 21st century is famous for a fast advancement in science and technologies. Describing the modern society, Florida (2002) refers to it as '3T approach'. He argues that talent, technology and tolerance are essential seeking to prosper in the 'creative age'. Technology and innovation are critical components of economic growth. Science is the basis of technology and innovation, and it is particularly important in today's creative society. "Yet in recent times fewer young people seem to be interested in science and technical subjects. Why is this? Does the problem lie in wider socio-cultural changes, and the ways in which young people in developed countries now live and wish to shape their lives? Or is it due to failings within science education itself?" (Osborne & Dillon, 2008, p. 5). Ohle et al. (2015) state that "Decreasing student interest and achievement during the transition from elementary to secondary school is an international problem, especially in science education" (p. 1211). It means that science educators encounter with challenging demands. Osborne and Dillon (2008) inquire about the motivation for learning science and claim that it is multidimensional in terms of sociological (new generation), psychological (new psychological motivation theories) and, educational issues (inquiry-based learning). In this monograph, the phenomenon of motivation is presented in the light of sociology, psychology, and education.

Motivation for learning Physics is rarely examined from the sociological perspective, which is based on the theory of generations. According to the sociological classification, persons born in 1995–2012 belong to Generation Z (McCrindle & Wolfinger, 2010; Seemiller & Grace, 2016). Currently, learners of Generation Z attend basic school. Is motivation for learning Physics of new generation exceptional? Are new generation students less interested in Physics than Generation Y learners? The problem of motivation of Generation Y learners for learning Physics is also acute: "A German learner at lower and upper secondary level regards Physics as very difficult to learn, very abstract and dominated by male learners. As a result, Physics at school continuously loses importance <…> We assume that knowing about the influence of motivation on learning Physics may lead to new insights in the design of classroom settings" (Fisher & Horstendal, 1997, p. 411). According to Fisher and Horstendal (1997), a very abstract content of Physics is one of the reasons for the reduced interest of Generation Y learners in Physics.

The new Generation Z learns using new technologies that can facilitate the absorption of the complex content. The relationship of Generation Z with technologies has been precisely defined by Cross-Bystrom (2010): "Generation Z *is* technology". The statement presupposes a very close relationship with technologies since the generation itself is equalled to technologies. The learners of this generation have lived in the world closely intertwined with technologies since early childhood (Cross-Bystrom, 2010). The Californian psychologist Rosen (2012) raises a question related to what teachers know about young people, who spend entire hours at the computer in different social networks. Rosen's question can be restated as follows: what do teachers know about the motivation of learners of Generation Z for studying science? (Peciuliauskiene, 2014).

The attractiveness of science subjects is topical for both education policymakers (OECD 2007; Science Education in Europe, 2011) and researchers (Juuti, Lavonen, Aksela, & Meisalo, 2009; Lavonen et al., 2008; Loukomies et al., 2013). The problem of the attractiveness of science is very wide, it is analysed from different aspects: the individualisation of learning (Zacharia & Olympiou, 2010), collaborative learning (Nedic, Machotka, & Nafalski, 2003), formation of the concepts of Physics (Bajpai, 2013), as well as the motivation for learning Physics (Changeiywo et al., 2011).

Solid theoretical background is needed for understanding the phenomenon of motivation. The phenomenon of motivation explained by Self-Determination Theory (SDT) is addressed in this monograph. The main idea of SDT is that humans are active and growth-oriented, seeking for the actualisation of their potentialities and fulfilling their basic psychological needs. These needs include autonomy, competency, and social relatedness (Ryan & Deci, 2002). They move the lives of learners in desired and specific directions rather than merely staying passive subjects. A person's motivation in a particular situation is a result of the interaction between immediate social context and the individual's need system that aims at fulfilment (Ryan & Deci, 2002; Vansteenkiste & Ryan, 2013).

The shift from the focus on motivation phenomenon to the focus on how to motivate students is prominent in the presented monograph. It is important to find learning activities meaningful and worthwhile, although they are not necessarily pleasurable per se for students. Claude Bernard, a famous 19th-century scientist, states that science is a "superb and dazzling hall, but one which may be reached only by passing through a long and ghastly kitchen" (cited in Osborne et al., 2003, p. 1074). It is our contention that inquiry is a good tool for successful "passing through a long and ghastly kitchen".

Edelson (2001) states that "in traditional science classrooms, content and inquiry skills are taught separately through separate learning activities. In these classrooms, content is taught didactically through lecture, reading, and problem sets, whereas scientific practices are taught through structured laboratory experiments" (p. 356). This traditional approach in education emphasises memorisation of facts and leads to 'inert knowledge' that cannot be called upon when it is useful (Whitehead, 1929). Inquiry-based learning allows overcoming the problem of 'inert knowledge' by describing how learning activities can foster useful conceptual understanding.

Scholars reveal that inquiry-based learning gives opportunities for students to develop positive attitudes towards science (Gormally, Brickman, Hallar, & Armstrong, 2009). While students engage in inquiry as a means, they are also supposed to learn scientific content knowledge through inquiry (Arnold, Kremer, & Mayer, 2014). Students learn how to do science and acquire relevant skills or abilities, as well as develop understanding of scientific inquiry itself (NRC, 1996).

Students learn more positively when teachers support their autonomy rather than control and pressurise them towards a specific way of thinking, feeling, or behaving (Vansteenkiste et al., 2004 Reeve et al., 2004; Reeve & Tseng, 2009a). Student engagement is a multidimensional construct consisting of four relatively equally weighted indicators: behavioural, emotional, cognitive, and voice (Reeve & Tseng, 2009b). Inquiry allows using an autonomy-supportive style in the classroom, which promotes student engagement as it supports an internal perceived locus of causality (PLOC), and a sense of choice in students (Reeve, Nix, & Hamm, 2003). When students engage in learning activities without the support of an internal locus, their engagement lacks the motivational foundation of personal interest. An autonomy supportive approach in science possesses inner motivational resources that are fully capable of directing their activity in Physics labs in more productive ways.

Well-constructed lab activities in Physics are powerful learning tools; through guided inquiry, students gain first-person experience of scientific principles and phenomena learned in lectures, and learn to employ experimental methods to solve discrete problems (Rissing & Cogan, 2009; Hughes & Ellefson, 2013).

A well-constructed Physics lab can ensure a high level of autonomy in the classroom. Autonomy-supportive teacher behaviour can be effective in fostering intrinsic motivation in learners (Reeve & Jang, 2006). Autonomy-supportive teacher behaviour can be supported by different levels of inquiry-based learning. Banchi and Bell (2008) identify four levels of inquiry-based learning: confirmation inquiry, structured inquiry, guided inquiry, and open enquiry. Low autonomy is acquired by confirmation inquiry, a higher autonomy is gained by structured inquiry and guided inquiry, whereas the highest autonomy is obtained by open enquiry. It means that the level of science labs can be related to motivation for learning science. Different environments (real, remote, digital) of science labs allow using different autonomy-supportive styles in the classroom, which can influence student engagement.

Competency and social relatedness focus the lives of students in desired and specific directions rather than making them stay as mere passive subjects (Ryan & Deci, 2002). Bybee and McCrae (2011) argue that students' interest in science topics decreases as the topic moves further away from personal experience and immediate relevance to students' own lives. Non-formal science activity approximates to students' personal experience and immediate relevance to their own lives. According to SDT, non-formal science activity allows fostering motivation for learning. How does this happen with students of new (Z) generation?

Students employ Mathematics to perform Physics labs. Mathematics is deeply woven into both teaching and practice of Physics (Redish, 2004; Redish & Gupta, 2010). Maths is treated as a language of Physics: "we explore Mathematics as a language and consider the language of Mathematics in Physics through the lens of cognitive linguistics" (Redish & Kuo, 2015, p. 561). According to Redish and Kuo (2015), Mathematics is deeply embedded in the practice of Physics; therefore, the Physics community continues to have mixed success of teaching students to use Mathematics effectively in Physics (Redish, 2004; Redish & Gupta, 2010; Redish & Kuo, 2015). "Other sciences, such as chemistry, biology, geology, and meteorology often use Mathematics extensively in practice, but typically rely less heavily on it than Physics does in high school and college instruction" (Redish & Kuo, 2015, p. 561). In the current monograph, the phenomenon of motivation is analysed focusing on Physics and Mathematics as ways of the deeper understanding of the phenomenon of motivation.

The discussed situation highlights the scientific problem of the present monograph, which is formulated as a question: how does inquiry-based learning affect the motivation of the new generation school students for learning Physics and Mathematics?

The object of the research is the motivation of school students for learning Physics and Mathematics.

The aim of the research is to reveal the effects of inquiry on the motivation of new generation students for learning Physics and Mathematics.

The objectives of the research are as follows:

1. How do real and digital Physics labs influence the motivation of new generation students for learning Physics?
2. What psychological student needs determined by SDT are assured by digital and real Physics labs?

3. How do Physics labs based on structured and guided inquiry engage school students in science?
4. How does the integration of responsible research and innovation in education engage school students in science?
5. How does informal Physics activity engage gifted students for learning Physics?
6. How does Maths language engage students for learning physics?

The Structure of the monograph

The aforesaid research objectives determine the content and structure of the monograph. In the first chapter of the monograph, *Theoretical background of new generation students' motivation for learning*, the phenomenon of motivation is analysed theoretically. The first section of this chapter, *Motivation for Learning: the Issue of Self-Determination Theory*, focuses on SDT, which explores the reasons why individuals engage in education activities. SDT is based on two perspectives: humans' basic needs for competency and self-determination. Intrinsic motivation is maintained only when humans feel competent and self-determined (Deci & Ryan, 1985). SDT consists of four sub-theories: Basic Needs Theory, Cognitive Evaluation Theory, Organismic Integration Theory, Causal Orientation Theory. The phenomenon of motivation for learning Physics and Mathematics is analysed in terms of diferent sub-theories.

The second section is entitled *Inquiry perspective into learning science*. The concept of inquiry is defined according to the *National Science Education Standards* (NRC, 1996), in which inquiry "involves making observations; posing questions; examining books and other sources of information to see what is already known in light of experimental evidence; using tools to gather, analyse, and interpret data; proposing answers, explanations, and predictions; and communicating the results. Inquiry requires identification of assumptions, use of critical and logical thinking, and consideration of alternative explanations" (NRC, 1996, p. 23). The quoted paragraph emphasises scientific inquiry, which is defined as the activities of exploring and investigation searching for answers to problem questions. Scientific inquiry can be implemented in educational practice by different instructional models. The 5E instructional model is described. Each of the 5 E's describes a phase of learning, and each phase begins with the letter 'E': Engage, Explore, Explain, Elaborate, and Evaluate. The 5E model starts with the Engage phase.

In the second section *Inquiry perspective into learning science*, the motivation for learning science is analysed in terms of inquiry levels. Our findings are interesting in the light of SDT focusing on the motivation of lower secondary school students for learning science at the labs based on structured and guided inquiry levels. Buck et al. (2008) relate the continuum of inquiry to the structure of continuum. Meanwhile, we link inquiry continuum to the basic psychological needs for competency, relatedness, and autonomy. A higher level of inquiry gives more autonomy but demands more competency and relatedness not only from the students of lab groups but also from other students in the virtual communities with scientists.

The third section, *Inquiry perspective on school students' motivation for learning science*, analyses motivation for learning science from the inquiry perspective. It was identified the essential inquiry components for motivation of learning: opportunities to control nature of learning, and collaboration. These components enhance learning autonomy and authenticity. Intrinsic motivation occurs only in activities that hold novelty, challenge, or aesthetic value for a student. It corresponds to inquiry levels: confirmation, structured, guided, and open inquiry. Students act as 'original researchers' at the levels of structured, guided, and open inquiry. They cannot find the answer to a research question only using the information obtained from literature and primary sources. They have to perform experiments in order to provide the answer to the research question. The activity of 'original researchers' enhances the autonomy and promotes the motivation of students for learning science.

The last section, *Where motivation for learning Math comes from: the issues of inquiry and SDT*, analyses motivation for learning Mathematics. To succeed in Physics, students need to be fluent in mathematical processing in the context of Physics. The present monograph focuses on the understanding of student motivation for learning Mathematics through the SDT, in particular Organismic Integration Theory (OIT).

In the second chapter of the current monograph *Methodology of motivation for learning Physics and Mathematics*, the philosophical commitments and learning theories of the research are introduced. The research builds on literature about learning theories: cognitivism, constructivism, and connectivism. Cognitivism focuses on the internal processes of learner, emphasises internal thought, and focuses on mental structures. It gives a background for internal and external loci of causality, which gives explanations of extrinsic and intrinsic motivation of learners, as well as the internalisation of the external locus.

The constructivist perspective asserts that learners construct knowledge from within, by being engaged in problem solving and by experiential learning. Since knowledge is actively constructed at Physics labs, learning is presented as a process of active discovery. The construction of new knowledge depends on the learning environment (hands-on, remote, digital labs) and learning method (structured, guided inquiry). Constructivists claim that motivation is both extrinsic and intrinsic in educational practice. On the one hand, learning is a social phenomenon; learners are partially motivated by rewards provided by the knowledge community. Moreover, knowledge is actively constructed by learners; learning also depends on the learner's internal drive to promote the learning process.

Connectivism is a learning theory of the digital age that emphasises the role of the cultural and social context in how and where learning occurs. Connectivism takes cognitivist principles one step further by asserting that learning does not simply happen within an individual, but within the networks. Connectivists argue that knowledge is distributed across a network of connections, and therefore, such learning consists of the ability to construct and traverse those networks (Downes, 2013). In the present monograph, the phenomenon of motivation is described at the level of responsible research and innovation activity, when learners communicate with other learners and scientists.

The third chapter of the monograph, *Empirical insights into the motivation of school students for learning science*, presents the data of empirical research. The structure of this chapter complies with the research objectives. In the first section, *Motivation for learning Physics at school: the case of different generations*, a comparative analysis of motivation of Generations Z and Y for learning Physics is provided. The comparative analysis is based on the secondary analysis of a data national research (2008).

The section *Digital and hands-on labs in Physics education at school: how do they engage?* focuses on the context of digital and hands-on labs in Physics education, and the potential of the teaching sequence for promoting students' motivation for learning. We analyse what psychological needs of learners determined by SDT are assured by digital and hands-on Physics labs.

The formation of intrinsic motivation for learning Physics is presented at different levels of inquiry: the cases of structured and coordinated inquiry. In the section *Formation of intrinsic motivation at different levels of inquiry*, an approach of inquiry levels is adopted, and justifications for the decisions made within it are scrutinised.

In the fourth section, *Socio-scientific issues and communication in science education* the motivating features of integrating responsible research and innovation into the practice of science education are investigated. This section is based on the theory of connectivism, which allows understanding how the communication with peers and scientists promotes motivation for learning science.

In the last section *Student motivation for participation in Physics competitions*, the role of non-formal Physics education in promoting motivation for learning Physics is analysed. The motivation of gifted students for learning Physics is analysed with reference to the SDT, namely its sub-theory Organismic Integration Theory (OIT).

All the sections of the chapter *Empirical insights into the motivation of school students for learning science* have the same structure: a short theoretical introduction, research methodology, and results.

The fourth chapter of the monograph, *Empirical insights into the motivation of school students for learning Mathematics*, considers the motivation for learning Mathematics. The data of a national research (2012) on the 8th grade students are analysed according to the model of motivation continuum from the lowest level of extrinsic motivation to intrinsic motivation. The trans-contextual model has been adapted to examine the effects of the perceived autonomy support and autonomous motivation of school students for Mathematics activities in the classroom on their autonomous motivation for doing Mathematics homework outside school.

The Discussion chapter combines the ideas presented in all the chapters of the monograph. The discussion includes the summary of the main findings of the monograph, followed by our interpretation of these results in the light of literature review presented in the introduction and the first chapter of the monograph. The results are examined and qualified, as well as inferences and conclusions are drawn.

The monograph's authorship is as follows: Pečiuliauskienė is the author of all the sections in the monograph except the section 3.5. *Student motivation for participation in Physics competitions.*

Aleksa is the author of the section 3.5 *Student motivation for participation in Physics competitions.*

1 Theoretical background of new generation students' motivation for learning

1.1 Motivation for learning: the issue of self-determination theory

By Palmira Pečiuliauskienė

The section focuses on Self-Determination Theory (SDT), which explores the reasons why individuals engage in education activities. According to SDT, it is important to fulfil three basic psychological needs of learners: the need for autonomy, the need for competence, and the need for relatedness. The article examines Physics education activity in the light of SDT.

The 21st century is famous for a fast advancement in science and technologies. Science is a part of European cultural heritage but "in recent times fewer young people seem to be interested in science and technical subjects. Why this?" (Osborne & Dillon, 2008). When searching for an answer to the question what factors determine a positive approach to the studies of natural sciences, first at all, we have to answer the question what motivation is. In scholarly literature, motivation is considered from various viewpoints. Therefore, the conceptualisation of motivation is a big challenge.

What does it mean to be motivated? The Latin root of the word *motivation* means *to move*. In this sense the study of motivation is the study of an action (Eccles & Wigfield, 2002). "To be motivated means *to be moved* to do something" (Rayan & Deci, 2000, p. 54). A person who feels no inspiration to act is, thus, characterised as unmotivated, whereas someone who is activated to the end is considered as motivated. Motivation is a theoretical concept used to explain behaviour. Theories of human motivation have evolved from the emphasis on reactive responses to pressures (external reinforcement contingencies or internally felt needs) to an emphasis on intrinsically motivated, self-determined actions (Pardee, 1990).

The concept of motivation is explained by different psychological theories. A number of motivation theories and their classifications are offered in scholarly literature. Eccles and Wigfield (2002) distinguish four groups of motivation theories: theories focusing on expectancy (self-efficacy theory and control theory), theories focusing on reasons (theories focused on intrinsic motivation, self-determination, flow, and goals), theories integrating expectancy and value constructs (attribution theory, the expectancy-value and self-worth theory), as well as theories integrating motivation and cognition (social-cognitive theories of self-regulation and motivation, theories of motivation and volition). The first focuses on beliefs about competency and expectancy for success. The second group of motivation theories deals with the reasons why individuals engage in different activities. These theories include constructs such as achievement values, intrinsic and extrinsic motivation, interests, and goals. The third group of theories integrate value constructs and expectancy. The fourth group explains the links between motivational and cognitive processes (Eccles & Wigfield, 2002).

Motivation refers to the process of goal-directed activities, which can have different reasons (Schunk, Pintrich, & Meece, 2008). One reason is to act from the perspective of individual actualisation of one's innate capacities for development. SDT is an example of this approach (Deci & Ryan, 1985, 2008, 2012b; Ryan & Deci, 2002). Another approach to motivation is the socio-cognitive view based on expectancy-value theory of motivation. This approach emphasises cognitive structures such as decision making and beliefs that occur in the presence of certain socio-contextual factors (Pintrich, 2003). The current monograph focuses on both perspectives: individual actualisation of one's innate capacities for development and the socio-cognitive view of motivation.

Analysing the motivation of new generation school students for learning Physics, a review of the works emerging from the SDT, which, according to Eccles and Wigfield (2002), belongs to the second group of motivation theories, is provided. SDT focuses on the reason why individuals get engaged in different science education activities. In the monograph, we focus on those that are most closely linked to the reasons why individuals get engaged in science learning activities. It means that we focus on SDT itself.

SDT proposed by Deci & Ryan (1985) is based on two perspectives on person motivation: persons have basic needs for competency and self-determination; persons 'are motivated to maintain an optimal level of stimulation' (Eccles &Wigfield, 2002, p. 112). In SDT, Deci and Ryan (1985) revealed different types of motivation based on the different goals that give rise to an action. SDT describes motivation as a continuum from amotivation to intrinsic motivation (Deci & Ryan, 2002). Intrinsic motivation is maintained only when humans feel competent and self-determined (Deci & Ryan, 1985). Intrinsic and extrinsic types of motivation have been widely studied, and the "distinction between them has shed important light on both developmental and educational practices" (Deci & Ryan, 2000, p. 54). An intrinsically motivated behaviour is characterised by engagement and concentration. It occurs spontaneously and people become wholly plunged in the activity (Byman et al., 2012). According to Loukomies (2013), intrinsically motivated individuals get freely engaged in certain activities with interest and enjoyment. Extrinsic motivation, on the other hand, is "a construct that pertains whenever an activity is done in order to attain some separable outcome" (Ryan & Deci, 2000, p. 60). This definition posits the difference between extrinsic and intrinsic motivation by explaining the reasons of the involvement in an activity.

SDT consists of four sub-theories: Basic Needs Theory, Cognitive Evaluation Theory (CET), Organismic Integration Theory (OIT), and Causal Orientation Theory (Deci & Ryan, 2012b). The phenomenon of motivation for learning Physics and Mathematics is analysed employing different sub-theories of the SDT in this monograph. Basic Needs sub-theory maintains that all human beings have fundamental psychological needs to be competent, autonomous, and related to others (Deci & Ryan, 2012b). The need for competency means that people desire to feel efficacious, to have an effect on one's environment, and to attain valued outcomes. The need for autonomy is expressed by a desire to be self-initiating acting with a sense of full endorsement and volition. According Ryan and Deci (2002), "to be autonomous does not mean to be independent". It means to have a sense of free will when doing something or acting out of own interests. The need for relatedness manifests itself by a desire to feel connected with and to be accepted by others. The need for relatedness deals with a desire to be connected to other people (Ryan & Deci, 2002).

CET is a sub-theory of SDT that is designed to explain the impact of external consequences on intrinsic motivation of humans. Two main dimensions of SDT are in the focus of CET – competency and autonomy. CET analyses how the need for autonomy and the need for competency of learners are affected by external forces. Deci (1975; cited in Vallerand & Reid, 1984) determines that positive feedback enhances intrinsic motivations and negative feedback diminishes it. Vallerand and Reid (1984) conclude that these effects are mediated by perceived control.

According to CET, intrinsic motivation is enhanced depending on whether the needs for autonomy and competency are supported. External events that promote greater perceived competency enhance intrinsic motivation, whereas those that diminish perceived competency decrease intrinsic motivation (Deci & Ryan, 1985). External events have three aspects: the informational aspect, the controlling aspect, and the amotivating aspect. According to Deci and Ryan (2012a), the informational aspect facilitates a perceived internal locus of causality and perceived competency, thus positively influencing intrinsic motivation (Fig. 1.1.1). The controlling aspect facilitates a perceived external locus of causality, thus negatively influencing intrinsic motivation. The amotivating aspect facilitates perceived incompetency, and undermines intrinsic motivation while promoting disinterest in the task (Deci & Ryan, 2012b).

Perceived locus of causality	Behaviour cause	Intrinsic motivation
External	Behavior cause lies outside human	Intrinsic motivation decreases
Internal	Behaviour cause lies inside the human	Intrinsic motivation increases

Fig. 1.1.1: The relationship between PLOC and intrinsic motivation.

The concept of perceived locus of causality (PLOC) was first introduced by Heider (1958). Heider (1958) distinguished two causations: personal causation - dependent from the person's intentions, and impersonal causation - independent from the person's intentions. There is a distinction between an internal PLOC (such as ability and effort) and external PLOC (behaviour is guided by external factors) (DeCharms, 1968). If a person has a stronger internal PLOC, they will feel they are in control of how they behave (*I can manage the situation*). If they have a stronger external PLOC, they will believe the environment has a greater influence over their activity (*I am dependent on the situation*). The distinction between internal and external PLOC has been crucial for the studies of intrinsic versus extrinsic motivation (Deci & Ryan, 1985). Considering Physics assignments, students evaluate them in terms of how well they meet their needs to feel competent and in control. The difficulty of Physics assignments is an external stable locus. Meanwhile, the success in performing them is an external unstable locus. Ability for Physics (stable) and motivation for learning Physics (unstable) are both internal locus. If a learner thinks they are able to complete the Physics assignment, they will be intrinsically motivated to complete it demanding no further external motivation (Fig. 1.1.1). It implies that intrinsic motivation for learning Physics will be enhanced or undermined depending on whether the needs for autonomy and competency are supported or thwarted in Physics education.

"To summarise, the CET aspect of SDT suggests that classroom and home environments can facilitate or forestall intrinsic motivation by supporting versus thwarting the needs for autonomy and competence. However, it is critical to remember that intrinsic motivation will occur only for activities that hold intrinsic interest for an individual – those that have the appeal of novelty, challenge, or aesthetic value for that individual. For activities that do not hold such appeal, the principles of CET do not apply" (Deci & Rayan, 2000, p. 49). In the monograph, students' activity in the project ENGAGE (Equipping the Next Generation for Active Engagement in Science) based on Responsible Research and Innovation

(RRI) is described. Students' beliefs in RRI activity are related to the progress of science. RRI considers that the progress of technology and science accounts for a better future. However, innovations must be planned carefully to address societal needs in accordance with societal values in order to maximise the benefits and reduce any harmful impact (Okada, 2016). Students' decision making is based on engaging, scientific dilemmas related to RRI. Decision making based on a scientific dilemma captures students' interest to discuss and extend their understanding, as well as develops their evidence-based opinion.

The second sub-theory of SDT is OIT. OIT has been introduced to explain different forms of extrinsic motivation and the contextual factors of motivation (Deci & Ryan, 1985). OIT argues that extrinsic motivation varies according to the internalisation ('how well the value of an activity is felt') and integration ('the process from external regulation to self-regulation') of the value of the activity (Ryan & Deci, 2000). Ryan and Deci (2000) define several levels of extrinsic motivation in the process of going from external to internalised regulation (Fig. 1.1.2). "These are external (regulation coming from outside the individual); introjected (internal regulation based on feelings that one has to do the behaviour); identified (internal regulation based on the utility of that behaviour (e.g., studying hard to get grades to get into college)); and integrated (regulation based on what the individual thinks is valuable and important to the self). Even this last level, however, is not fully internalised and self-determined" (Eccles & Wigfield, 2002, p. 113).

The main idea of OIT is that people have a natural inclination to internalise their experiences. Internalisation is a continuous process leading from amotivation to intrinsic motivation. Internalisation in the situations of learning Physics can be understood as seeing the value and importance of a Physics assignment by the student, whereas integration into this situation would be the degree of performing a Physics assignment as his/her own choice. As students are externally motivated and rewarded for an uninteresting Physics assignment, they will develop a certain degree of intrinsic motivation around it in the course of time. It means that external regulations can be internalised and become internal regulations. In an amotivating situation, students do not have any intention to perform a Physics assignment, or if they do, they do it in a passive way.

The feelings of autonomy are absent in the degree of external regulation and the external rewards will not be internalised at all (Gagné & Deci, 2005). According to Ryan and Deci (2000), such behaviours are performed only by seeking to satisfy an external demand or obtain an external reward contingency. Let us consider one example of external regulation: *I learn Physics well because my parents promise to buy me a new bike for good results in Physics.*

Behavior	Nonself Determined →					Self Determined
Type of Motivation	Amotivation	Extrinsic Motivation				Intrinsic Motivation
Type of Regulation	Non-Regulation	External Regulation	Introjected Regulation	Indentified Regulation	Integrated Regulation	Intrinsic Regulation
Locus of Causality	Impersonal	External	Somewhat External	Somewhat Internal	Internal	Internal

Fig. 1.1.2: The continuum of motivations and regulations (simplified model of Ryan & Deci, 2000, p. 61).

Introjected regulation is a type of behaviour, which is performed in order to avoid the sense of shame or guilt, and is ultimately "based in contingent self-esteem" (Ryan & Deci, 2002). Introjected regulation provides some level of autonomy, as well as external regulation is internalised to some degree. Introjected regulation is ego involvement, with the focus on the approval of oneself or others. Here is an example of introjected regulation: *I do Physics tasks because I want my classmates to keep thinking I am good at Physics.*

According to Ryan and Deci (2000), a more autonomous, or self-determined, form of extrinsic motivation is regulation through *identification*. The person identifies personal importance of behaviour, and thus, accepts its regulation as his or her own. At the level of identified regulation, humans begin to "accept the underlying value of behaviour" (Ryan & Deci, 2000). Humans may experience a degree of autonomy and competency. More importantly, at this stage regulation begins to come in line with socially reinforced values (*I do Physics tasks well because I know the knowledge of Physics is valuable for me*).

The most autonomous form of extrinsic motivation is *integrated regulation.* Integration occurs when identified regulations is fully assimilated by the self (*I do Physics tasks because I am a physicist; the competency in Physics is part of my professional competency*). The more humans internalise the reasons for an action and assimilate them by themselves, the more one's extrinsically motivated actions become self-determined (Ryan & Deci, 2000).

Intrinsic motivation is a prototype of self-determined activity. At this level extrinsic regulations become transformed into intrinsic motivation. According to Deci and Ryan (2012b), *intrinsic motivation* refers to doing something because it is interesting or enjoyable. Intrinsic motivation differs from extrinsic because in the case of extrinsic motivation the extrinsic goal is self-endorsed. The component of relatedness is important in fostering intrinsic motivation, since intrinsic motivation exists in the intercourse of individuals and activities. Interpersonal events (e.g., rewards, feedback) that conduce towards *feelings of competency* during the action can enhance intrinsic motivation. Events and structures will *not* enhance intrinsic motivation unless they are accompanied by "*a sense of autonomy* or, by an *internal perceived locus of causality*" (DeCharms, 1968). People must experience satisfaction of the needs both for competency and autonomy aiming at a high level of intrinsic motivation. The authors of the current monograph analyse how the needs for competency, autonomy, and relatedness influence intrinsic motivation for learning Physics.

Causality Orientations Theory (COT) is the fourth sub-theory of the SDT. COT focuses on the differences that exist between humans and their intrinsic or extrinsic motivators. The main precondition for COT is the assumption that not everyone is equally motivated by either extrinsic or intrinsic regulators. Not everyone requires the same sense of competency or autonomy (Weinstein, Przybylski, & Ryan, 2012). To describe this, COT offers three orientations: autonomy orientation (a result of satisfying the basic needs); controlled orientation (a result of satisfying the needs for competency and relatedness); and impersonal (a result of satisfying the needs for competency, relatedness, and autonomy). Behaviour that is more autonomous is associated with more creative learning and engagement (Roth, Assor, Kanat-Maymon, & Kaplan, 2007). In the present monograph, motivation has not been analysed from the aspect of COT. This can be considered as a research limitation. On the other hand, this opens a door to far-reaching research on the role of COT in student motivation.

1.2 Inquiry perspective into learning science

By Palmira Pečiuliauskienė

The section deals with the concept of inquiry-based learning. A higher level of inquiry gives more autonomy but demands more competency and relatedness. The section emphasises scientific inquiry, which is defined as the activities of exploration and investigation searching for answers to questions.

In the first chapter of the monograph, the phenomenon of motivation for learning according to SDT is presented. According to SDT, intrinsically motivated behaviour is characterised by engagement and concentration. Intrinsic learning motivation describes the quality of motivation during the process of learning and means the "doing of an activity for its inherent satisfactions rather than for some separable consequence" (Deci & Ryan, 2000, p. 56). Intrinsically motivated students enjoy learning and want to learn more about the content (Jurik, Gröschner, & Seidel, 2014). It occurs spontaneously and people become wholly plunged in the activity (Byman et al., 2012). Loukomies (2013) states that intrinsically motivated individuals engage in certain activities freely, directed by feelings of interest and enjoyment. According to SDT, intrinsic motivation is fostered "when we provide people [with the] opportunity to satisfy their basic psychological needs for competency, relatedness and autonomy" (Deci & Ryan, 1991, p. 329). People have a need to feel competent, autonomous, and affiliated with others when doing a task (Shumow et al., 2013).

Brophy (2008) proposes to shift from focusing on intrinsic motivation to focusing on how to motivate students to learn, i.e. to find learning activities meaningful and worthwhile even though they are not necessarily perceived as pleasurable per se by students. Loukomies et al. (2013) support Brophy's ideas and provide further evidence that, irrespective of the context or country, students learn science with regard to what they view as meaningful and valuable activities. Engaging students into personally meaningful activities in a socially supportive environment is a strong motivation to produce valuable work and to get encouraged for higher order learning (Pintrich et al., 1994; Ryan & Patrick, 2001; Lawlor et al., 2015).

Over the past two decades, the interest in the role of inquiry in science education has increased (NRC, 1996; Chiappetta, 1997; NRC, 2000; Rocard et al., 2007). Inquiry corresponds to the inductive approach in education, the so-called bottom-up approach, which gives space for observation, experimentation and teacher-guided construction of students' knowledge (Rocard et al., 2007). Inductive approach is opposed to deductive, or 'top-down transmission', approach, where learners, being passive receivers of knowledge, are forced to handle abstract notions (Rocard et al., 2007).

Chiappetta (1997) defines inquiry-based learning of science in the following way: "Upon entering a science classroom, one should be able to observe an exciting learning environment in which students are wondering why and finding out. Students should be asking questions, resolving discrepancies, figuring out patterns, representing ideas, discussing information, and solving problems. This vision of science teaching is associated with the term inquiry" (p. 24). Authors of the NRC (1996) provide a similar definition (the definition is quoted in the Introduction of this monograph). Chiappetta identifies two distinct meanings of inquiry: general inquiry and scientific inquiry (Chiappetta, 1997). General inquiry can be thought of as learning through discovery to find out everything about a specific topic. Students are offered experiences in which they can develop an understanding of science concepts, as well as connect concepts and ideas with the phenomena experienced in everyday life. Scientific inquiry stresses active student learning and the importance of understanding a specific topic.

Colburn (2000) states that "Perhaps the most confusing thing about inquiry is its definition. The term is used to describe both *teaching* and *doing* science" (Colburn, 2000, p. 42). The concept *inquiry* is used ubiquitously both as a style of teaching and as a method for conducting research (Flick, 1995). Brown et al. (2006) describe this dilemma and claim that "What makes this research difficult to understand is the lack of agreement about what constitutes an inquiry-based approach" (p. 786).

Other researchers (Anderson, 2002; Minner et al., 2010) describe three distinct meanings of inquiry: scientific inquiry, inquiry learning, and inquiry teaching. According to Llewellyn (2011), general inquiry can be referred to as an activity which is based on asking questions to seek information, seeking and collecting evidence, confirming or rejecting suspicions, and communicating the outcome of the question. "Scientific inquiry can be defined as activities of explorations and investigations in order to answers a problem question, which can be posed by the teacher, the textbook, or the students themselves" (Llewellyn, 2011).

According Riga et al. (2017), science and inquiry should go hand in hand. "Scientific inquiry refers to the diverse ways in which scientists study the natural world and propose explanations based on the evidence derived from their work. Inquiry also refers to the activities of students in which they develop knowledge and understanding of scientific ideas, as well as an understanding of how scientists study the natural world" (National Research Council, 1996, p. 23).

General inquiry skills can be developed by scientific inquiry (Edelson et al., 1999) and scientific practices (NRC, 2012): posing and refining investigation questions, planning and managing investigations, gathering facts, exploring possibilities, conducting research, thinking through discoveries, as well as analysing and communicating results (Alisinanoglu et al., 2012). It means that inquiry provides an opportunity to acquire specific investigation skills, relying on different methods of investigation and different sources of evidence. Inquiry activities should capture cognitive processes of authentic science (Chinn & Malhotra, 2002; Reiff, Harwood, & Phillipson, 2002). According to Harwood (2004b), in the process of inquiry, scientists move among these activities in unique paths and repeat activities as often as they find necessary. An activity model of the scientific inquiry process contains 10 activities (with possing questions) in which scientists are engaged (Fig. 1.2.1) (Reiff, Harwood, & Phillipson, 2002). First, inquiry activity focuses on "Ask questions" activities, since asking general questions is the core feature of scientific inquiry (Fig. 1.2.1). Divergent questions are the key feature of any scientific inquiry. Reiff, Harwood, & Phillipson (2002) state that the following nine activities must be done to carry out an inquiry: define the problem, form the question, investigate the known, articulate an expectation (may be a formal hypothesis), carry out the study, examine the results, reflect on the findings, communicate with others, and make observations (Fig. 1.2.1).

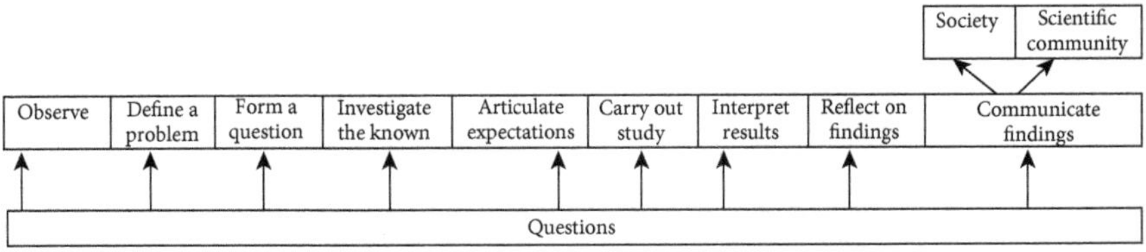

Fig. 1.2.1: The activity model for the process of scientific inquiry (simplified model of Reiff, Harwood, & Phillipson, 2002).

In scientific inquiry, research activities must be done often in classroom. Harwood (2004a) analyses scientific inquiry of the students in biology labs, organised by Levri and Levri (2003). The researcher identifies 16 points in time, where Levri and Levri's students move from one activity to another (Fig. 1.2.2). Analysing this map, Harwood (2004b) maintains that this model can help science teachers and science teacher educators in framing inquiry teaching and learning. Physics labs is a good example of scientific inquiry because learners have a possibility to pose and refine investigation questions, to plan labs, to gather facts, to explore possibilities, to think through discoveries and to analyse results, as well as to communicate.

Scientific inquiry can be implemented in educational practice by different instructional models. According to Jobrack (2011), an instructional model, or learning cycle, is a sequence of stages teachers may go through to help students develop a full understanding of a lesson concept.

The 5E instructional model, developed by Bybee in the 1980s, was specifically designed to provide a model that supports a constructivist approach (Bybee, 2002). Constructivism promotes the idea that people construct knowledge and meaning from their experiences. Vygotsky has formulated the main principle of constructivism referred to as the "zone of proximal development" (Vygotsky, 1978). This zone marks the distance between the actual developmental level that an individual is capable of, and the potential the individual acquires under an adult's guidance or with the help of capable peers.

Each of the 5 E's describes a phase of learning, and each phase begins with the letter "E": Engage, Explore, Explain, Elaborate, and Evaluate. The 5E model starts with the *Engage* phase. At this phase, students should make connections between past and present learning experiences in their activity. They should become mentally engaged in the concept, process, or skill to be learned. At the *Explore* phase, students actively explore their environment or manipulate materials, as well as identify and develop concepts, processes, and skills. At the third *Explain* phase,

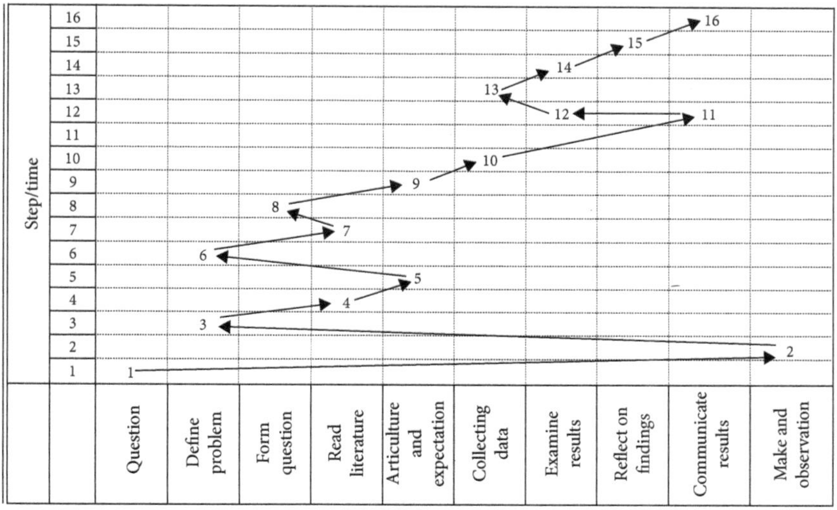

Fig. 1.2.2: Chronological map of activities students are engaged in during a lab exercise (Harwood, 2004a).

students explain the concepts they have been exploring. They have opportunities to verbalize their conceptual understanding or to demonstrate new skills or behaviours. At this phase, teachers can introduce formal definitions and terms. At the fourth 5E phase, students develop a deeper and broader understanding of major concepts, obtain more information about the areas of interest, and refine their skills. At the last phase students assess their understanding and abilities. They assess the result of the 'zone of proximal development' and let teachers evaluate their understanding of key concepts and skill development.

In this monograph, scientific inquiry is described on the basis of the project ENGAGE, which focuses on the impact of communication (Fig. 1.2.1) at Physics labs on students' motivation for learning science carrying out Physics labs on the 5E model. Scientific inquiry must be approached by formal and informal education. A five-step method was implemented to coordinate the formal and informal learning in the ENGAGE project. The project started with formal activity in the classroom, which encompassed three steps: engage, explore, and explain (Tab. 1.2.1). The fourth step of the project (*expand*) (inter-student relations, teacher–student relations to scientists) was carried out in informal activity. The last step of students' project activity was implemented by formal classroom activity.

The ENGAGE project (duration 2014–2016) is a practical example of the promotion of students' motivation by scientific inquiry in different countries.

Tab. 1.2.1: Scientific inquiry in terms of the 5E model: the issue of the ENGAGE project.

Stage	Purpose	Content	Activity
Engage	Interest	Introduction of a problem Students can propose the problem based on images/video clip, discussion of initial ideas/class vote	Formal in a classroom
Explore	Understanding	Reminder of science knowledge and introducing the RRI process Engaging activity (game) where students experience the process and understand the principals involved, introduction of thinking guide – breaking RRI skill into chunks	
Explain	Process	Use process they have learnt in Explore to discuss the problem set up in Engage Apply science content, use of *thinking guide*, use of *argumentation tool* (claim/ evidence/debate)	
Extend	Practice	Use the RRI skill to consider another problem in a different, but similar context. Opportunity to discuss how RRI skill applies to other contexts	Informal outside a classroom
Evaluate	Evaluate learning	Assess understanding of RRI skill and science knowledge Present a justified answer to the RRI problem	Formal in a classroom

The theoretical background of the ENGAGE project is based on SDT which highlights the fulfilment of the three basic psychological needs: the need for autonomy, the need for competency, and the need for relatedness (Okada et al., 2015). Addressing the scientific inquiry tool, the 5E instructional model has been employed to show how to use it to teach an enquiry process and apply science content in a contemporary issue (Boddy et al., 2003; Bybee et al., 2006), using the materials created during the ENGAGE project. The 5E model aims at deriving the concepts and principles from scientific investigations and experimentation, so that apart from acquiring knowledge, students develop skills such as critical reasoning and problem solving. In the 5E model, the curriculum is designed to allow students to explore scientific phenomena and their own ideas. Students are invited to explain their ideas, and explanations can also be provided by a teacher, a textbook, or through the usage of technologies. Students are able to explore, explain, extend, and evaluate their own progress. During the inquiry-based activity, students analyse the process and product of innovations and achieve acceptable and even desirable outcomes (Stahl, 2013). The content on big science and technology impact has been taken from the ENGAGE project site. This content gets students engaged in the possible consequences of research and innovation, as well as expands communication from inter-student relations to teacher–student relationships with scientists (Fig. 1.2.1, Communicate findings).

According to Buck et al. (2008), a common goal for science educators is to engage students in inquiry. However, many factors complicate the completion of such a task. Summarising the conclusions of the research on inquiry, Colburn (2000) states that inquiry requires deductive reasoning and concrete thinkers have a difficulty developing an understanding of abstract concepts. For example, inquiry-based instruction is great for showing students that the evaporation process depends on the temperature of the environment. This is a concrete idea that students can investigate while accomplishing Physics labs. However, inquiry-based methods are less effective in helping students understand how scientists explain the phenomena through the kinetic-molecular theory.

Students cannot be expected to immediately be able to plan and carry out their own research. They develop their inquiry abilities and understandings about the possibility to do the research. It is obviously that students' inquiry abilities depend on their age and inquiry practice. The more familiar the activity, materials, and context of the investigation are, the easier it is for students to learn through inquiry. Banchi and Bell (2008) state that "Luckily, there are many levels of inquiry that students can progress through as they move toward deeper scientific thinking" (p. 26). Only two questions remain – what are the levels of inquiry? On what basis can they be distinguished?

Some scholars suggest distinguishing four levels of inquiry: confirmation, structured, guided, and open inquiry (Bell, Smetana, & Binns, 2005; Banchi & Bell, 2008). The levels of inquiry depend on guiding questions, procedure, and expected results (Tab. 1.2.2). Banchi and Bell (2008) focus on how much information is provided to students and how much guidance you provide as the teacher.

At the first level, *confirmation inquiry*, students are provided with the question and procedure (method). The results of investigation are also known in advance. For example, you may want students to confirm that the all bodies in a vacuum fall with the same acceleration. Students know that a freely falling object has a constant acceleration of 9.8 meters per second squared. Students follow the directions for doing the experiment, record their data, and analyse their results.

Tab. 1.2.2: The four levels of inquiry and the information given to the student in each one (Banchi & Bell, 2008, p. 27)

Inquiry level	Question	Procedure	Solution
Confirmation inquiry	+	+	+
Structured inquiry	+	+	
Guided inquiry	+		
Open inquiry	+		

At the second level, *structured inquiry*, the question and procedure are still provided by the teacher but the results of the investigation are not known in advance. For example, students need to collect data showing that all bodies acquire a constant acceleration of 9.8 meters per second squared in a vacuum. According to Banchi and Bell (2008), confirmation and structured inquiry are considered lower-level inquiries, and they are very common in elementary science curricula. These levels are important because they enable students to gradually develop their abilities to conduct a more open-ended inquiry.

In *guided inquiry*, students know only the research question. Students design the procedure (method) to test their research question and the resulting explanations. They have had numerous prior opportunities to learn and practice different ways to plan experiments and record data.

The fourth and highest level of inquiry is *open inquiry*. Students have the best "opportunities to act like scientists, deriving questions, designing and carrying out investigations, and communicating their results. This level requires the most scientific reasoning and greatest cognitive demand from students" (Banchi and& Bell, 2008, p. 27).

We have compared inquiry levels by Banchi and Bell (2008) and inquiry approaches by Chiappetta (1997): general inquiry and scientific inquiry (Tab. 1.2.3). Confirmation inquiry and structured inquiry correspond to the approaches of general inquiry, whereas guided inquiry and open inquiry correspond to scientific inquiry. (Tab. 1.2.3). The scientific approach requires the most scientific reasoning and greatest cognitive abilities from students. Students experience at general inquiry will be able to successfully conduct guided or open inquiries.

In this monograph, the motivation for learning sciences has been analysed at the levels of structured and guided inquiry. The middle levels of inquiry have been chosen because it has been decided to compare the impact of general inquiry and scientific inquiry on learning science. Lower secondary school provides a possibility to do structured and guided inquiry. Open inquiry is used only in scientific Physics projects accomplished in informal Physics education.

Tab. 1.2.3: The comparison of four levels of inquiry with general and scientific inquiry

Inquiry level	Question	Procedure	Solution	Inquiry approaches
Confirmation inquiry	+	+	+	General inquiry
Structured inquiry	+	+		General inquiry
Guided inquiry	+			Scientific inquiry
Open inquiry	+			Scientific inquiry

Individuals with different motivation orientations develop motivation for learning science through being engaged in activities that may fulfil their basic psychological needs, yet different aspects of activity appeal differently to different students (Loukomies et al., 2013). Autonomy-supportive teacher behaviour can be effective in fostering intrinsic motivation in students (Reeve & Jang, 2006; Reeve & Lee, 2014; Jang et al., 2016). It can be supported by different levels of inquiry-based learning. Low autonomy is acquired by confirmation inquiry, a higher autonomy is gained by structured inquiry and guided inquiry, whereas the highest autonomy is obtained by open enquiry. Our monograph deals with the application of structured and guided inquiry in Physics labs using real and digital devices, which provide different autonomy of experimental activity.

Inquiry levels are related to learning theories, which attempt to describe how people learn. When we understand how students learn, we can provide instructions that make learning more engaging and efficient. There are three basic learning theories that have been proposed by educators and psychologists: Constructivism, Cognitivism, and Behaviourism. Inquiry levels have been developed in response to each learning theory. Confirmation inquiry is more related to Cognitivism. Other levels of inquiry (structured inquiry, guided inquiry, open inquiry) are more related on the theory of Constructivism. The current monograph is restricted to the exploration of inquiry based on Constructivism and Cognitivism. The impact of Physics labs corresponding to structured and guided inquiry on the motivation of students has been described.

We can envision the theoretical background of Inquiry levels towards motivation for learning in the Inquiry level models (Banchi & Bell, 2008; Buck et al., 2008). According to SDT, intrinsic motivation is fostered when we provide opportunity for students to satisfy their basic psychological needs for competency, relatedness, and autonomy (Deci & Ryan, 1991). The models of inquiry levels are related to one psychological need, i.e. the need for autonomy. However, the authors of inquiry levels do not use the concept of autonomy. They employ the concept of guidance instead. Buck et al. (2008) state: "*Inquiry and the National Science Education Standards* (NRC, 2000) presents inquiry as a continuum, and

Tab. 1.2.4: A rubric to characterize inquiry in the laboratory (Buck et al., 2008, p. 54)

Characteristics	Level 0: Confirmation	Level ½: Structured inquiry	Level 1: Guided inquiry	Level 2: Open inquiry	Level 3: Authentic inquiry
Problem/ Question	Provided	Provided	Provided	Provided	Not provided
Theory/ Background	Provided	Provided	Provided	Provided	Not provided
Procedures/ Design	Provided	Provided	Provided	Not provided	Not provided
Results analysis	Provided	Provided	Not provided	Not provided	Not provided
Results communication	Provided	Not provided	Not provided	Not provided	Not provided
Conclusions	Provided	Not provided	Not provided	Not provided	Not provided
More structure ◄					► Less structure

Brown et al. (2006) extrapolates this continuum with a figure moving from more to less guidance" (p. 52). Less guidance means more autonomy and independence in students' activity in science labs. "For each experiment or activity, we analysed each characteristic based upon the criterion of student independence" (Buck et al., 2008, p. 53). According to Buck et al. (2008), students have more independence when less information about lab activity is provided for them (Tab. 1.2.4). The six characteristics represent areas in lab activity where students can act autonomously: problem/question; theory/background; procedures/ design; analysis of results; communication of results; conclusions provided (Tab. 1.2.4) (Buck et al., 2008, p. 54).

Buck et al. (2008) state that the *problem/question* characteristics refers to the topic of investigation in the activity. Students can formulate the question under investigation themselves or it is provided in the lab text. Students have more autonomy when they formulate the question under investigation. According to SDT, more autonomy allows fostering students' motivation for learning.

Theory/background refers to all prior knowledge necessary for the investigation. The *procedures/design* specifies the lab activity that students perform. The *results analysis* characteristics refers to how data are interpreted and analysed. *Communication of results* specifies the manner by which data and experimental results are presented, and whether the options of how to communicate the results of lab activity are provided or not. If the manual provides a summary or

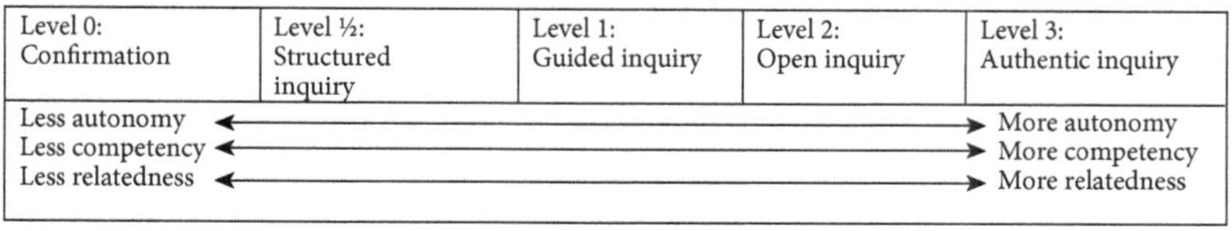

Fig. 1.2.3: Inquiry levels and continuums of the basic psychological needs for competency, relatedness and autonomy.

a list of results that should have been obtained in the lab activity, it means that *conclusions* are provided and vice versa (Buck et al., 2008, p. 54).

Structured inquiry corresponds to Level ½. Following the classification of Buck et al. (2008), at the level of structured inquiry the laboratory manual (or the teacher) provides the problem, procedures, and analysis by which students can discover relationships or draw conclusions that are not known from the manual. Guided inquiry corresponds to Level 1. The laboratory manual (or the teacher) highlights the problem and procedures, yet the methods of analysis, communication, and conclusions are for the students to discover themselves.

Our findings are interesting in the light of SDT of motivation of lower secondary school students for learning science in the labs organised at the levels of structured and guided inquiry. Buck et al. (2008) associate inquiry continuum with structure continuum (Tab. 1.2.4). We link inquiry continuum with basic psychological needs for competency, relatedness, and autonomy (Fig. 1.2.3). A higher level of inquiry gives more autonomy but demands more competency and relatedness not only from the students of lab groups, but also from other students in virtual communities with scientists. In our monograph, we are searching for the solution how structured and guided inquiry fulfils the basic psychological needs for competency, relatedness, and autonomy in real and virtual Physics labs.

1.3 Inquiry perspective on school students' motivation for learning science

By Palmira Pečiuliauskienė

The section chapter deals with motivation for learning science in terms of inquiry levels. The chapter reveals that students can assume different roles at different levels of inquiry. Students act as 'information synthesisers' at the level of confirmation inquiry and act as 'original researchers' at the levels of structured, guided, and open inquiry.

People's healthy tendencies toward growth and integrity can be explained by SDT constructed by Deci and Ryan (2002). The main idea of SDT is that humans are active and growth-oriented, seeking for the actualisation of their potentialities and fulfilling their basic psychological needs. These needs include *autonomy, competency, and social relatedness*. They move the lives of learners in desired and specific directions rather than being passive subjects. A person's motivation in a particular situation is a result of the interaction between immediate social context and the individual's need system that seeks fulfilment (Ryan & Deci, 2004; Vansteenkiste & Ryan, 2013).

Motivation for learning sciences from the perspective of inquiry is not new. Ormerod and Duckworth (1975) emphasises the phenomenon 'swing from science' and highlights the disaffection with science and technology among students. Osborne et al. (2003) states that "While the body of research conducted has been good at identifying a problem, it has had little to say definitively about how the problem might be remediated" (p. 1073). Sholars render a proposal how to 'swing to science' by identifying the essential components of motivation: opportunities to choose, control nature of learning, and collaboration. Students prefer more opportunities for practical work in science, as well as opportunities for discussion – all of which provide an enhanced role for personal autonomy and learning authenticity (Osborne et al., 2003).

Inquiry-based science learning is based on the idea that science learning should be authentic (Edelson et al., 1999). Authenticity in learning contexts becomes a stimulation for motivating learners to acquire new knowledge and actively engage in the learning sequence (Greeno, Collins, & Resnick, 1996). It corresponds with the ideas of Self-Determination sub-theory CET, which suggests that classroom and home environments can facilitate or forestall intrinsic motivation. Intrinsic motivation will occur only for activities that hold novelty, challenge, or aesthetic value for that individual (Rayan & Deci, 2000).

Scholars reveal that inquiry-based learning gives opportunities for students to develop positive attitudes towards science (Gormally, Brickman, Hallar, & Armstrong, 2009). While students engage in inquiry as a means, they are supposed to also learn scientific content knowledge through inquiry (Arnold, Kremer, & Mayer, 2014). Students learn how to do science and acquire relevant skills or abilities, and they develop an understanding of scientific inquiry itself (NRC, 1996). According to Rissing and Cogan (2009), the inquiry approach allows students to connect classroom activities with their personal experiences and students are more motivated to learn. The study of Rissing and Cogan (2009) proves that students are more motivated when they participate in an inquiry laboratory.

Students learn better when teachers support their autonomy rather than control and exert pressure on them for a specific way of thinking, feeling, or behaving (Vansteenkiste et al., 2004; Reeve et al., 2004; Reeve & Jang, 2009). Student engagement is a multidimensional construct consisting of four relatively equally weighted indicators: behavioural, emotional, cognitive, and voice (Reeve & Tseng, 2009a). Inquiry allows using an autonomy-supportive style in the classroom, which promotes student engagement because it supports students' internal PLOC, experience of volition, and a sense of choice (Reeve, Nix, & Hamm, 2003). When students are engaged in learning activities without the support of an internal locus, their engagement lacks the motivational foundation of personal interest. An autonomy-supportive approach in science possesses inner motivational resources that are fully capable of directing their science lab activity in productive ways.

Etkina et al. (2010) describe the results of Physics labs accomplished by undergraduate students with different level of autonomy in non-design and design Physics labs. According to Etkina et al. (2010), non-design labs are more or less traditional. During design labs, students have to design their own experiments. Students are provided with lab hand-out questions that focus students' attention on the elements of the scientific process inquiry: representing the situation, deciding on the experiment, and analysing experimental uncertainties (Etkina et al., 2010). Design labs are an example of scientific inquiry. Solving an experimental problem, a scientist needs to decide which features can be ignored and which features are important, how to represent the problem in different ways, how to use available equipment to collect necessary data; how to evaluate the quality of the measurements; and how to share the results. "For example, if the goal of the experiment was to determine the coefficient of friction between their shoe and the floor, and they obtained a certain value through the experiment and calculation, they could not ask the lab instructor whether the value was acceptable. Instead, they had to design an independent experiment to determine the same value and then make a judgment about the results of both experiments" (Etkina et al., 2010, p. 66). It means that at non-design Physics labs students have less autonomy, whereas at design labs students have a possibility to act more independently. The level of inquiry in non-design labs corresponds to structured inquiry (problem, procedures, and analysis by which students can draw conclusions that are known in advance). Meanwhile, design labs correspond to the level of guided inquiry (the methods of analysis, communication, and conclusions are for the students to design). Hence, design labs give more autonomy for undergraduate students.

It has been proved that when students are engaged in the design of experiments, they do not only develop scientific abilities (write lab reports that receive much higher scores on all scientific abilities), but also use them without prompts and scaffolding (Etkina et al., 2010). This conclusion confirms the idea that design labs correspondent to guided inquiry are more engaging to ungraduated students.

Inquiry levels can be analysed in terms of information literacy. Exner (2014) analyses information literacy of scientist (faculty, doctoral students, and other research-based graduate students). The activity of scientists corresponds to the activity of scientific inquiry. According to Exner (2014), there are fundamental differences between the processes of inquiry used to find answers to a research question. She states that there are two different ways of searching for information in scientific inquiry. "Most searchers locate existing information and synthesise it to form a conclusion. These information synthesizers can make an assumption that the components of their answer exist; combine existing knowledge to answer a question. But original researchers cannot ever answer their question by searching. <…> The original research searcher is preparing for an empirical investigation that will answer a question that is original, unique, and unanswered" (Exner, 2014, p. 460). It means that empirical inquiry is main component which separates the activity of 'information synthesizers' from 'original researchers'. In empirical inquiry, students have to refine a question not only from the literature (Analytical cycle with the literature), but also from experimental activity (Analytical cycle with empirical findings). An 'original researcher' may need to use the existing information and analysis to lead to experimentation, as well as additional analysis and original thought before a final answer is reached (Exner, 2014 p. 461).

In the process of information synthesis, students can give an answer to a research question by searching for information in literature and primary sources. According to Exner (2014), during this process students have to get a general topic area in the class, make a preliminary statement with subtopics, explore information to learn more, and then refine main statement.

Students can act like 'information synthesizers' or like 'original researchers' in Physics labs. It corresponds to inquiry levels: confirmation, structured, guided, and open inquiry (Bell, Smetana, & Binns, 2005; Banchi & Bell, 2008). Higher levels of inquiry are more based on empirical findings (Fig. 1.3.1). At the levels of structured, guided, or open inquiry, students cannot get the answer to a research question only by searching for information in literature and primary sources (labs textbooks). They have to perform an experiment in order to provide the answer. Exner (2014) does not analyse the process of information synthesis in term of learner motivation. On the basis of theoretical analysis made in this

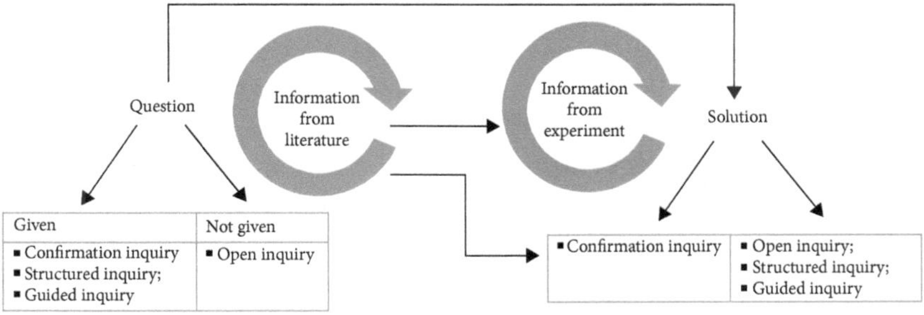

Fig. 1.3.1: Inquiry levels and analytical cycles at Physics labs.

section and on the analytical cycle of information synthesis process, we can conclude that two cycles of information synthesis process get students more engaged in sciences as they are given more autonomy.

The analytical cycle with empirical findings depends on the lab environment. Laboratory experimentation is generally regarded as an important element of science education (Lowe et al., 2013). An early survey of students' attitudes towards science and environmental variable is provided by Gardner (1975). It is a focused research with a strong emphasis on the laboratory or experiential elements of science learning (Freedman, 1997; Hofstein & Lunetta, 2004; Etkina et al. 2010). Lowe et al. (2013) state that there is a lot of work over an extended period on students' attitudes towards science, but only limited research towards laboratories exists.

A large body of research in this area focuses on comparing different lab environments – hands-on, remote, and virtual laboratories. Physics labs can be real or virtual. There are various ways of defining virtual labs. They can be defined as computer programmes that allow student to run simulated experiments via the web or as stand-alone applications. A virtual lab could be a set of simulations put together (e.g. applets, flash base demos, animations) (Bajpai, 2013). Scientific literature reveals many features of virtual labs: tools are less time-consuming, more flexible, clean, rapid, safe, and they open up such types of experimentation that otherwise might not be possible for students to engage in (Dalgarno & Lee, 2010; Petersson, Andersson, & Säljö, 2013).

There is a discussion about the role of real and virtual labs in education. It is argued that there is no simple answer to the dilemma which laboratory is best for students. All types of laboratories offer certain advantages. The balanced mixture of real, virtual and remote labs should be offered for students (Nedic, Machotka, & Nafalski, 2003), both forms of activity have unique properties that are needed to promote a deeper conceptual understanding (Winn et al., 2006; Jaakkola & Nurmi, 2008). The authors (Jaakkola & Nurmi, 2008) suggest that simulations can be used first to help students understand the underlying theoretical principles, and that laboratory exercises can then be used to demonstrate that those principles apply in the real world.

Bhargava et al. (2006) note that virtual labs reduce the need for equipment, are available at any time from any place, provide more information to students, and offer students an opportunity to work at their own pace while exploring difficult or interesting sections. Thornton (2008) points out that an activity-based or research-based environment supporting peer learning is the best environment for student learning in Physics. Darrah et al. (2014) have tested how well a laboratory (real or virtual) supplements the teaching of Physics concepts done in lectures and elsewhere. The analysis shows no evidence that one of the treatments (virtual or hands-on) was more effective than the other in conveying the concepts of the labs to students. "We conclude that the Virtual Physics Lab software used in these two introductory physics courses produced similar learning outcomes as the hands-on traditional labs" (Darrah et al., 2014, p. 812).

In recent years, a large number of virtual labs have been produced. Research shows that modern interactive laboratory experiments increase motivation when students enter interactive laboratories from mobile phone/computer/etc. (Zacharia & Olympiou, 2010; Changeiywo, Wambugu, & Wachanga, 2011; Oidov, Tortogtokh, & Purevdagva, 2012). Virtual environment is attractive if operations carried out in virtual labs by some students are observed by other students (Ince et al., 2014). Virtual labs are attractive because students can easily use working methods that resemble methodologies practiced by scientists such as observation, developing hypotheses, explaining observations, and testing these hypotheses using given datasets (Petersson, Andersson, & Säljö, 2013).

Remote web-based labs are a new environment in science education. Remote labs have a lot of advantages over the traditional science lab format: cost saving, access to labs by more students (disabled students that are marginalised from regular science work) and others (Lnag, 2012). Lnag (2012) notices that there is an ongoing debate about the merits of remote web-based science labs in comparison to other possible types of labs, as well as there is disproportionately little research evaluating their effectiveness. Lnag (2012) compares students' results in conducting a Physics lab using remote web-based technology and their results while using traditional labs. On the basis of the conducted research, Lnag (2012) states that there "was no indication that one format was superior in general" (p. 13).

The same conclusion is drawn by Lowe et al. (2013). There is no significant difference between the educational outcomes for students who performed an experiment remotely, versus those who carried out a hands-on experiment (Lowe et al., 2013). Lowe et al. (2013) state that "online, remote-access, pre constructed experimental apparatus is less engaging than hands-on practical work in school learning spaces, such as laboratories <...> that the remote laboratory experience requires less interpersonal negotiation and social interaction and a narrower selection of science-oriented skills than does hands-on experimentation" (p. 1216). This conclusion does not correspond to the constructivist learning ideas.

The constructivist learning environment should provide for interaction between learners, promote higher-level learning and social presence, and to help develop personal meaning (Tam, 2000; Heinich et al., 2002). The collaboration and cooperation among learners should enhance their interaction, as well as create learning communities to negotiate and co-construct meaning (Neo & Neo, 2009). In this monograph, the impact of lab environment on the engagement of school students for learning sciences is analysed. We reveal the role of digital and hands-on labs (real) in engaging students in sciences.

Psychological relatedness is one of the three main intrinsic needs involved in self-determination. It is any relationship between two or more individuals. There is a gap in literature that considers intrapersonal indicators of collaboration as ways of motivation (Lotrecchiano et al., 2016). There is also a gap in the line of research inquiry in the team science literature in terms of individual and intrapersonal indicators for participation in team, such as interpersonal and technological aspects of scientific collaboration (Stokols et al., 2008a; Stokols et al., 2008b; National Academy of Science, 2015). Compared to other types of teams, science teams contain unique aspects caused by the expectations from the knowledge-generating environment in which they operate (National Academy of Science, 2015). It has become a priority for behavioural and social scientists to better understand the underlying factors supporting or deterring individual engagement in collaborative research (Fiore, 2008). Team and management scientists have worked to understand the motivational peculiarities in several teaming environments, such as sports or management teams (Ehrlinger et al., 2005; Hurtz and Williams, 2009). Only a few inquiries focus on motivational peculiarities specific to scientists and engineers using an expectancy model (Goodman et al., 1970; Lotrecchiano et al., 2016).

Scholars point out that researchers are motivated to collaborate with others because of enjoyment of collaborating (Maglaughlin & Sonnenwald, 2005). Alon (2010) maintains that the satisfaction and enjoyment of team members as having a strong effect on innovation. Lotrecchiano et al. (2016) analyse motivators of communication in scientist teams. They distinguish six domains of motivation and threats to collaboration in scientist teams: resource acquisition, advancing science, building relationships, knowledge transfer, recognition and reward, and maintenance of beliefs.

Students work in groups and act like collaborative knowledge producing teams in structured and guided inquiry science labs. In this monograph, the statistical relationship between engagement of students and their relatedness in structured and guided Physics labs is explored. However, we do not focus on motivation and threats to collaboration in Physics lab teams according to the six domains (resource acquisition, advancing science, building relationships, knowledge transfer, recognition and reward, and maintenance of beliefs).

1.4 Where motivation for learning Math comes from: the issues of inquiry and SDT

By Palmira Pečiuliauskienė

In this section, Mathematics is treated as the language of Physics. The chapter focuses on the understanding of student motivation for learning Mathematics and Physics through SDT and inquiry-based learning.

The relationship between Physics and Mathematics is an object of the research conducted by philosophers, physicists, mathematicians, and educators (Uhden et al., 2011). Nowadays STEM (Science, Technology, Engineering, and Mathematics) background joins Science and Mathematics into one system. In order to foster positive attitudes and promote interest and motivation, many educational innovations (context-based teaching, inquiry-based teaching, and others) have been proposed, both in Science and Mathematics education (Fortus, 2014).

The relation between Mathematics (Maths) achievement and Physics study results is reported by a number of researchers. The low achievement of students in Physics is caused by low achievement in Mathematics (Ben, 2010; Olusola & Rotimi, 2012). A positive correlation between Mathematics abilities and Physics results has been detected (Meltzer, 2002). On the basis of analysis of discussion data results, Wenno (2015) concludes that "there is a positive and significant relation between both of interest and knowledge of Mathematics basic concept with student's ability to solve Physics problems" (p. 5). According to Zusho et al. (2003), self-efficacy plays an important role in improving mathematical problem-solving skills of lower secondary school students and science achievement in high schools.

Redish & Kuo (2015) state that "we explore maths as a language and consider the language of maths in physics through the lens of cognitive linguistics" (p. 561).

Mathematics provides symbols expressed in Physics to show the relation structure between different factors (Long & Jiar, 2014). Mathematics in Physics is not the same as doing Mathematics itself. Mathematics in Physics has a different purpose – to represent the meaning of Physical systems rather than to express abstract relationships (Redish & Kuo, 2015). According to Redish & Kuo (2015), "Learning math in math class and math in physics class should be treated as learning two related but distinct languages: Although there is significant overlap, there are also important differences, and expertise in one does not guarantee expertise in another" (p. 587). Redish & Kuo (2015) argue that Mathematics and Physics overlap but have different languages. We pose the following questions: do the psychological needs for motivation of learning Mathematics and Physics overlap or are they different? What is the role of inquiry in Mathematics educational practice?

Motivation for learning Physics depends not only on internal (interest, knowledge, abilities), but also on external (teaching method by teachers, educational subject) factors (Changeiywo et al., 2011). Emotions and motivation are important for learning Mathematics and sciences (Schukajlow et al., 2017). We are interested what the peculiarities of the motivation of new generation students for learning Mathematics and Science are. In this chapter of the monograph, we are looking for an answer to the question: are inquiry and SDT powerful issues in the educational practice of Mathematics?

Scholars highlight basic psychological needs (autonomy, competency, and relatedness) and focus on the understanding of student motivation and self-regulated learning in Mathematics and Science in relation to perceived teacher autonomy support (Ng et al., 2015). The findings of these scholars (Ng et al., 2015) offer a perspective to secondary students' basic psychological needs in the contexts of Mathematics and Science. On the basis of SDT (Deci & Ryan, 1985), Ng et al. (2015) state that "competency involves experiencing optimal self-proficiency", "autonomy refers to being the source of one's behaviour", whereas "relatedness refers to a sense of belonging to other individuals of the community" (p. 1362).

Describing the need of autonomy in Mathematics educational practice, Ng et al. (2015) quote Reeve (2002), stating that "autonomy support refers to identifying and fostering students' intrinsic motivation by giving them choices or opportunities; fostering understanding and interest with respect to learning; providing rationales and informational feedback; as well as encouraging self-regulated learning" (Reeve, 2002, p. 185).

Ng et al. (2015) argue that students having good motivated strategies for learning Mathematics seem to perceive competency as a more prevalent need than autonomy. "Perceived competency seemed to be the most pertinent need across the four groups, with the greatest effect size among three psychological needs" (p. 1371). Consequently, teachers should provide opportunities to solve challenging tasks to the students with a strong feeling of competency.

Scholars consider inquiry-based learning a powerful tool for the promotion of motivation for learning Maths and Science. In inquiry-based Maths education, students produce, discover, and evaluate scientific knowledge (Kim, Tan, & Talaue, 2013). Inquiry-based learning allows fostering scientific thinking and reasoning skills; students are encouraged to experience the process of knowledge construction (Chinn & Malhotra, 2002). "In this approach, students are likely strategic and engaged in learning contexts such as mathematics and science" (Ng et al., 2015, p. 1361).

Savelsbergh et al. (2016) state that "Many teaching approaches have been tried to improve student attitudes and achievement in science and mathematics education. Achievement effects have been synthesized, but a systematic overview of attitude effects is missing" (p. 158).

Other scholars notice that one of the most prominent aspects is that 30 years of research into teaching approaches on attitudes in Science and Mathematics education "has [have] been bedeviled by a lack of clarity about the concept under investigation" (Osborn et al., 2003, p. 1053).

Over the last decade, several review studies have documented the effects of innovative teaching (using information and communication technology , inquiry-based learning, collaborative work, out-of-school activities) and found a modest positive effect on motivation for learning science (Potvin & Hasni, 2014). The evidence in Mathematics seems to be sparser (Savelsbergh et al., 2016). Inquiry-based learning in Mathematics is analysed by Rosen (2009). The researcher reports a positive impact of 'problem posing' on achievement and mo tivation for learning Mathematics (Rosen, 2009).

Summarizing the survey review about science and mathematics education from 1988 to 2014, Savelsbergh et al. (2016) state: "To sum up, although the general tendency across studies seems to be that innovative teaching approaches do have positive effects, there is little clarity about what interventions cause effects on what outcome, and under what conditions" (p. 159). In order to shed more light on what interventions cause motivation of school students for learning Physics and Mathematics, inquiry-based learning is analysed in this monograph.

Aiming at comparing the effectiveness of interventions and synthesis of the findings across studies as we have done in our review, we have chosen inquiry and SDT as conceptual frameworks. To answer the research question about the influence of inquiry-based learning on school students' motivation for learning Physics, two inquiry levels (structured and guided) have been selected. A trans-contextual model has been adapted to examine the effects of school students' perceived autonomy support and autonomous motivation for Mathematics activities in the classroom on their autonomous motivation for Mathematics homework outside school.

Summary of the theoretical background

Modern views on inquiry-based learning can provide innovative alternatives for the educational practice of science at school. In the present monograph, inquiry-based learning is analysed in terms of motivation at different inquiry levels (Tab. 1.2.3), as well as different inquiry environments (Fig. 1.3.1). The conceptualisation of motivation is based on the SDT, which highlight three basic psychological needs: the need for competency, the need for autonomy, and the need for relatedness. According to Ryan and Deci (2002), motivated behaviour may be self-determined or controlled. Self-determined or autonomous behaviour arises from one's self (Deci & Ryan, 1985). To be autonomous in science education means to study with a full sense of volition. Motivation affects two primary cognitive processes: PLOC and change in perceived competency (Ryan & Deci, 2002).

In the well-organised educational practice of science, PLOC changes towards a more internally perceived locus, thus enhancing intrinsic motivation. Intrinsically motivated students perceive the locus to be within themselves.

Autonomy-supportive teacher behaviour can be effective in fostering intrinsic motivation in learners (Reeve & Jang, 2006). Autonomy-supportive teacher behaviour can be supported by different levels of inquiry-based learning. H. Banchi and R. Bell (2008) identify four levels of inquiry-based learning: confirmation inquiry, structured inquiry, guided inquiry, and open enquiry. The lowest level of inquiry (confirmation inquiry) corresponds to activities where learners know the possible outcomes of a project, and where a detailed description of activities and problems is provided. The second level of inquiry (structured inquiry) is reached in projects when learners are provided with a problem and the method for its solution. The third level (guided inquiry) is characterised by the fact that learners know the problem but have to find out how to solve it by themselves. The highest level (open inquiry) is attained when learners identify a problem, methods for its solution, and explanations for the cross-curricular phenomena themselves. Low autonomy is acquired by confirmation inquiry, a higher autonomy is gained by structured inquiry and guided inquiry, whereas the highest autonomy is obtained by open enquiry. The monograph deals with the application of structured and guided inquiry in Physics labs in terms of autonomy and intrinsic motivation in Physics labs activity. We have taken into account the environments (real, digital) in which Physics labs are carried out. We have investigated the construct of the intrinsic motivation of school students in real and digital labs.

Scholars of psychology initially described intrinsic and extrinsic motivation as a dichotomy. Later, the idea of the antagonistic nature of intrinsic and extrinsic motivation was limited (Deci & Ryan, 1985). Extrinsically motivated behaviours can become increasingly autonomous through the internalisation process.

OIT, a sub-theory of the SDT, describes four types of extrinsic motivation based on how self-determined or autonomous each type is, namely: external, introjected, identified, and integrated forms of regulation. In the present monograph, we analyse the types of extrinsic motivation on the basis of the secondary analysis of the data of the national research on Maths education. According to OIT, behaviour regulation may be autonomous (self-determined) or controlled, depending on the degree of internalisation. Intrinsically motivated individuals get freely engaged in certain activities, led by the feelings of interest and enjoyment (Ryan & Deci, 2002). Intrinsic motivation arises from a desire to learn a topic due to its inherent interests, self-fulfilment, enjoyment, and achievement of mastery of the subject (Ryan & Deci, 2009). Chapter four of the monograph *Empirical Insights into the Motivation of Lower Secondary School Students for learning Mathematics* is intended to consider motivation for learning Maths. *The data of national research (2012) of 8th grade students have been analysed according to the OIT model of motivation continuum from the lowest levels of extrinsic motivation to intrinsic motivation.*

2 Methodology of motivation for learning Physics and Mathematics

2.1 Basic paradigms of motivation for learning

By Palmira Pečiuliauskienė

In this part of the monograph the philosophical commitments and learning theories of the research are introduced. The research builds on literature about learning theories: cognitivism, constructivism, and connectivism.

The research of this monograph is based on philosophical commitments and learning theories. There are many different educational philosophies that can be applied as the background of educational research. Our research on motivation for learning Physics and Mathematics is based on the positivist, pragmatic, and progressivist paradigm.

Positivism confirms the role of science and Mathematics in the process of cognition. The fundamental basis of the paradigm of positivism is: objective reality exists that can be known and, thus, reality is inherently ordered. All knowledge regarding the matters of fact is based on the data of experience and beyond the realm of fact is pure logic and Mathematics.

Pragmatism has also tremendously influenced our research, as it emphases action in educational practice. The central theme of pragmatism is activity. Pragmatists state that the truth can be known only through its practical action and see thought as intrinsically connected with action. According to the pragmatic paradigm, teaching methods have to focus on hands-on activity, experimenting, and projects. Students should apply their knowledge to real situations through experimental inquiry. In this monograph, scientific inquiry is analysed on the basis of John Dewey's ideas on scientific inquiry. Dewey does not prioritise the empirical method over other methods but argues that all scientific methods sprout from the same logic (Loukomies, 2013). According to Dewey, subjects that are difficult and cause the learner to struggle should be organised and designed to build motivation about the topics. Student should enjoy learning and leave with a sense of accomplishment (Agarwal et al., 2010). In the current monograph, we are searching for the answer how inquiry-based learning facilitates such difficult subjects as Physics and Mathematics. We focus on experimental activity at Physics and Maths education.

A systematic search for motivation of new generation students for learning Physics and Maths is impossible without the progressivist paradigm. Progressive educators are associated with Dewey's 'child-centred' approach. Progressivists state that education should focus on the whole child, rather than on the content or the teacher. Although Dewey has stressed the needs of an individual child, the progressive movement diverges from his work with its emphasis on the child's immediate gratification and freedom of expression. Students should test ideas by active experimentation. Learning is based on the questions of learners that arise through experiencing the environment. Structured and guided inquiry makes it possible to test ideas, to learn through questions, to learn from one's individual experience in the physical and socio-cultural context.

Structured and guided inquiry has a lot of traits of progressivism: emphasis on hands-on projects, integrated curriculum focusing on thematic units, group work and development of social skills, collaborative projects, as well as education for social responsibility.

Pragmatism and progressivism serve as theoretical backgrounds for modern learning theories. As a form of pragmatism, constructivism focuses on how learners arrive at the knowledge that enables them to cope with the world (Good, Wandersee, & Julien, 1993). Constructivism is a theory of knowledge in the making – a theory of how learners make sense of the world (Good, Wandersee, & Julien, 1993). Von Glaserfeld (1993) refers constructivism as a theory of knowing, as opposed to the theory of knowledge.

Several learning theories are usually classified as constructivist: situated cognition, activity theory, experiential learning, anchored instruction, and authentic learning (Mattar, 2010). Our research bears inherent traits of these theories, the theory of constructivism in particular. This theory emphasises the importance of action on the part of the learner. Learning is considered as an active construction process, not a passive reception of knowledge.

Constructivist perspective asserts that learners construct knowledge from within, by getting engaged in problem solving, as well as by experiential learning. "Learners construct new knowledge and understanding based on what they already know and believe, and organize it in ways that facilitate retrieval and application" (Loucomies, 2013, p. 11).

Two forms of constructivism are proposed by Piaget and Vygotsky. Jean Piaget emphasises the importance of the cognitive processes that occur within individuals - 'cognitive constructivist' (Piaget, 1978). Cognitive constructivism, therefore, emphasizes personal construction of knowledge. Lev Vygotsky highlights the importance of society, culture, and language - 'social constructivism' (Vygotsky, 1978). Social constructivism emphases communication when students are given opportunities to interact with adults and peers in order to negotiate meaning (Palmer, 2005).

Constructivism is the dominant paradigm of learning in science. "When students are learning about science they use their existing knowledge, beliefs, interests, and goals to interpret any new information, and this may result in their ideas becoming modified or revised" (Palmer, 2005, p. 1854). Since knowledge is actively constructed at Physics labs, learning is presented as a process of active discovery. The construction of new knowledge depends on the learning environment (hands-on, remote, and digital labs) and learning method (structured and guided inquiry). In this monograph, we analyse the phenomenon of motivation for learning Physics and Mathematics from cognitive constructivist and social constructivist perspective. We analyse motivation for learning Physics at the levels of structured and guided inquiry from the cognitive constructivist aspect. The impact of responsible research and innovations in Physics educational practice is analysed from the aspect of social constructivism.

Motivation is recognised as an important factor in the construction of knowledge in the classroom. Constructivists assert that motivation in educational practice is both extrinsic and intrinsic. On the one hand, learning is a social phenomenon; learners are partially motivated by rewards provided by the knowledge community. On the other hand, knowledge is actively constructed by learners; learning also depends on the learners' internal drive to promote the learning process. Constructivist view on motivation is related to the main ideas of different sub-theories of the Self-Determination Theory.

Constructivists assert that motivation is intrinsic in the educational practice. According to Basic Needs sub-theory, learners' internal drive to promote the learning process is successful when their fundamental psychological needs to be competent, autonomous, and related to others are fulfilled. Constructivists maintain that motivation is extrinsic in the educational practice. The main idea of Organismic Integration Theory (OIT) is that people have a natural inclination to internalise their experiences. Internalisation is a continuous process leading from amotivation to intrinsic motivation. Individuals do not need a stimulus to be passively involved into action, since they maintain a dynamic relationship with the environment. Physical interaction with the environment is the basis for knowing.

Tam (2000) relates constructivism to the construction of technology-supported learning environments.

"Learning theories should be adjusted in a time in which knowledge is no longer acquired in linear manner, technology performs many of the cognitive operations previously performed by learners" (Mattar, 2010, p. 10).

Digital environments provide a possibility to infuse constructivist principles where learners are expected to be self-motivated, interactive, and collaborative participants in their learning experiences (Mattar, 2010). Connectivism is a learning theory of the digital age that emphasises the role of cultural and social context in how and where learning occurs. Connectivism is closely related to situational cognition theory of constructivism. Wilson and Myers (2000) state that situational cognition does not only mean concrete learning, but emphasises a network of social systems. Connectivism employs cognitivist principles and move one step further by asserting that learning does not simply happen within an individual, but within and across the networks. According to Siemens (2004), technological development and social software significantly alter the way we access information, and dialogue with each other.

"Learning is no longer a process that is entirely under the control of the individual, an internal, individualistic activity: it is also outside of ourselves, within other people, an organization or a database, and these external connections, which potentiate what we can learn, are more important than our current state of knowing" (Mattar, 2010, p. 10).

Connectivists argue that knowledge is distributed across a network of connections, and therefore, such learning consists of an ability to construct and traverse those networks (Downes, 2007). In the current monograph, we describe the phenomenon of motivation in responsible research and innovation activity, where learners communicate with others, as well as with scientists. In structured and guided inquiry, students have a possibility to learn across the networks looking for an answer to the main research question.

2.2 Research methods and instruments of motivation for learning

By Palmira Pečiuliauskienė

This part of the monograph focuses on the methods and instruments of the research, as well as their use and construction.

It introduces the underlying principles of instruments, intending to gather useful data. Our research is based on the quantitative approach, which is a systematic empirical investigation of motivation for learning phenomena via statistical or computational techniques. The objective of quantitative research is to develop mathematical models pertaining to motivation for learning Physics and Maths phenomena.

The process of measurement constitutes the basis for quantitative research, as it provides the fundamental connection between empirical observation and mathematical relationships.

The process of measurement depends on the process of operationalization. Operationalization means "translating a very general research aim or purpose into specific, concrete questions to which specific, concrete answers can be given. The process moves from the general to the particular, from the abstract to the concrete" (Cohen, Manion, & Morrison, 2007, p. 81). The general aim of our research is based on the SDT, which gives a theoretical background for motivation phenomena.

The Intrinsic Motivation Inventory (IMI) is a measurement device intended to assess learners' subjective experience related to the target activity in laboratory experiments. IMI has been used in several experiments related to intrinsic motivation and self-regulation (Ryan, Koestner, & Deci, 1991). The instrument consists of seven subscales. IMI assesses learners' interest/enjoyment, perceived competency, effort, value/usefulness, felt pressure and tension, perceived choice, and social relatedness while performing a Physics lab. There is only one subscale that assesses intrinsic motivation, namely the subscale of interest/enjoyment, although the overall questionnaire is entitled the IMI. The IMI questionnaire encompasses a negative predictor of intrinsic motivation - pressure/tension. *Effort* is a separate variable that is relevant to some motivation questions, the *value/usefulness* subscale is used in internalisation studies (Deci et al., 1994). We received permission to use IMI questionnaire from Edward Deci.

We used the IMI questionnaire in order to reveal the impact of the environment (hands-on, digital) on intrinsic motivation of learners (Section 3.2: *Digital and hands-on labs in Physics education at school: how do they engage?*); in order to reveal the impact of inquiry level (structured, guided) on intrinsic motivation of learners (Section 3.3: *Formation of intrinsic motivation at different levels of inquiry*); in order to reveal impact of responsible research and innovations (Section 3.4: *Socio-scientific issues in science education*).

Researchers have tested the reliability and validity of the IMI (McAuley et al., 1989; Tsigilis & Theodosiou, 2003; Eow et al., 2010). Their results generally demonstrate acceptable reliability and validity of the IMI with their respective populations and tasks. The reliability of IMI with Lithuanian school students has been checked in assessing intrinsic motivation gained through the engagement with responsible research and innovation (Cronbach Alpha value 0.829); in different environments (Cronbach Alpha value 0.738), and at different inquiry levels (Cronbach Alpha value 0.755).

The Lithuanian translation was conducted in an iterative process. First, we translated the questionnaire into Lithuanian. Then another researcher translated the Lithuanian version back to English. The original and the translated English versions were compared, and required revisions were conducted. The items within subscales were randomly ordered in the questionnaire, and the word order of some questions was reversed.

IMI variables were measured by a five-point Likert scale (1 = 'item in my case is not at all true' through 5 = 'item in my case very true'). According to Cohen, Manion, and Morrison (2007), ordinal scales, like Likert's, are in the subtlety of the responses and measurements developed for interval scales, and may be employed for ordinal Likert scale measurements.

In order to explore how inquiry-based learning and e-learning determines the motivation for learning Physics of learners of Generation Y, the secondary correlative analysis of the results of national research has been conducted. National research of the achievements of tenth formers was carried out in 2004 and in 2008.

In order to explore how inquiry-based learning determines the motivation for learning Mathematics, the secondary correlative analysis of the results of national research has been conducted. The data of the national research (2012) on 8th grade students are analysed according to the model of motivation continuum. One group of questions (M1) concerning the motivation for learning Mathematics is included into the Questionnaire about Mathematics for 8th grade students (Questionnaire about mathematics for 8th grade students, 2012). Exploratory factor analysis of questions of group M1 has been conducted.

The qualitative research was used to analyse motivation of tenth formers for learning Physics and its determinant factors. A semi-structured interview revealed the attitude of tenth formers towards the accomplishment of virtual and real laboratory work, as well as its role for motivation to study Physics. The results of qualitative research are described in Section 3.1: *Motivation for learning Physics at school: the case of different generations.*

Empirical part of our monograph consists of six different researches. We describe methods, instruments, sampling and samples of research more precisely in each empirical part.

3 Empirical insights into the motivation of school students for learning science

3.1 Motivation for learning Physics at school: the case of different generations

By Palmira Pečiuliauskienė

The chapter deals with e-learning, the relationship of virtual and real activities with motivation of tenth formers of generations Y and Z for learning Physics. The secondary correlative analysis of the results of the national research (2004 and 2008) has been conducted. The data reveal the impact of inquiry-based and computer-related activity on the motivation for learning Physics of Y generation learners. The qualitative research reveals the attitude of Z generation students towards real and digital Physics labs.

3.1.1 Introduction

The phenomenon of inquiry-based learning is impacted by changing generations. According to the sociologists' classification, persons born in 1977–1994 belong to Generation Y whereas the ones born in 1995–2012 belong to Generation Z (McCrindle & Wolfinger, 2010). Currently learners of Generation Z attend comprehensive school. The relationship of Generation Z with technologies has been precisely defined by A. Cross-Bystrom (2010): "Generation Z *is* technology".

Not much educational research has been performed on the motivation to learn Physics of preceding (e.g. Generation Y) and current generation (Generation Z) students as well as on the role of virtual and laboratory work to enhance motivation for learning Physics. Current researches are focusing on presenting new technologies and disclosing their advantages rather than analysing the impact of these technologies on learning process and its outcomes (Furberg, 2010).

The problem of motivation to study Physics was also important for Generation Y: "German students at lower and upper secondary level regard Physics as very difficult to learn, very abstract and dominated by male students. As a result, Physics at school continuously loses importance <...> We assume that knowing about the influence of motivation on learning Physics may lead to new insights in the design of classroom settings" (Fisher & Horstendal, 1997, p. 411).

The topics that are relevant to students' lives foster intrinsic motivation (Brozo, 2005). Autonomy-supportive teacher behaviour can be effective in fostering intrinsic motivation in students (Reeve & Jang, 2006). Suitable level of task difficulty is important for intrinsic motivation (Margolis & McCabe, 2006). The autonomy in the learning process plays a positive role for the developing of intrinsic motivation (Reeve & Jang, 2006). The different level of physics labs (confirmation inquiry, structured inquiry, guided inquiry, and open enquiry) (Tab. 1.2.2) provides different level of learning activities autonomy.

The physics labs can be real or virtual. Virtual labs can be defined as a computer program that allows student to run simulated experiments via the web or as a stand-alone application (Bajpai, 2013). The scientific literature reveals many features of virtual labs: tools are less time-consuming, more flexible, clean, rapid, safe, and that they open up for types of experimentation that otherwise might not be possible for students to engage in (Dalgarno & Lee, 2010; Petersson, Andersson, & Säljö, 2013). Research shows that modern interactive laboratory experiments increase motivation when students enter interactive laboratories from mobile phone/computer/etc. (Zacharia & Olympiou, 2010; Changeiywo, Wambugu, & Wachanga, 2011; Oidov, Tortogtokh, & Purevdagva, 2012). The virtual labs are attractive because students can easily use working methods that resemble methodologies practiced by scientists such as observation, developing hypotheses, explaining observations, and testing these hypotheses using given datasets (Petersson, Andersson, & Säljö, 2013; Ince, Kirbaslar, Yolcu et al., 2014). The discussed situation highlights *a scientific problem* which is formulated as a question: how do e-learning and inquiry-based learning impact the motivation for learning Physics of the learners of Generations Y and Z?

The object of the research is the motivation for learning Physics of learners.

The aim of the research is to reveal the statistical relationship between inquiry and the motivations' variables for learning Physics of learners of Generations Y and Z.

The objectives of the research are as follows:

1. How do real physical experiments activity is related with the motivation for learning Physics of Y Generation leaners?
2. How does e-learning is related with the motivation for learning Physics of Y Generation leaners?
3. What is the attitude of Z generation leaners towards real and virtual physical labs and their role for motivation to learn Physics?

3.1.2 Method of research

In order to explore how inquiry-based learning and e-learning determine the motivation for learning Physics of learners of Generation Y, the secondary correlative analysis of the results of national research has been conducted. National

Tab. 3.1.1: Number of respondents participating in national research (2004–2008)

Year	Number of tenth-form respondents	Date of birth
2004	3083	1984–1987
2008	3217	1990–1993

research of the achievements of tenth formers was carried out in 2004, and 2008. The research sample is reliable (Tab. 3.1.1) and representative (probability cluster sample). The research tools are also valid and reliable. The scope of our research is limited to the data of 2004 and 2008. The data of 2004 is covering learners born in 1984–1987 with ones born in 1987 composing the major part (75 %). According to sociological theory, learners born in 1984–1987 belong to Generation Y (1977–1994). Learners covered by the data of 2008 were born in 1990–1993 (Tab. 3.1.1). The major part of these respondents is born in 1991 (80.6 %). Thus they also represent Generation Y.

National research provides data on the learner attitudes towards learning Physics and their motivation to learn this subject. Extrinsic motivation is related to the assessment performed by the teacher (Question B34e) (National student achievement test, 2008).

To explore intrinsic motivation, particular research questions have been selected. The main selected question was: *What is your attitude towards Physics – Do you like Physics?* (Question GF1) (National student achievement test, 2008). Motivation was explored with regard to experiential learning and the questionnaire for learners included a number of questions related to structured-coordinated activities of an inquiry in classes of Natural Sciences. Such type of questions in 2004 questionnaire included asking the learners to define how often they *a) searched for data or information in texts, tables, diagrams, or charts; b) explained the causes of phenomena; c) applied the knowledge of natural sciences to deal with everyday issues* (Question GF7) in Physics classes. Besides, performing laboratory work in Physics was also explored: *a) you plan and conduct laboratory research or experiments with the help of a teacher; b) you employ different devices and tools* (Question GF8) (National student achievement test, 2004). The teacher-assisted planning of physical laboratory work, search for information in different sources as well as explanation of the obtained data are characteristic features of coordinated inquiry.

Considering this, the following factors leading to motivation for learning Physics were distinguished: planning of inquiry-based activity, search for information in traditional and non-traditional information sources, explanation of the causes of analysed phenomena, drawing of conclusions, and knowledge application in everyday situations.

The surveys (2004 and 2008) also comprised questions about the application of information and communication technologies (ICT). Only one question was provided in the general survey for learners in 2004 (B30): *how often do you do the following: a) play computer games; b) send and read e-mails;* and *c) use the computer for learning* (National student achievement test, 2004). The survey of 2008 included more questions related to ICT application: *a) what do you use computer for after school: a) for playing games; b) for learning; c) for browsing on the internet; d) for communication (e-mails, skype, chats, etc.); e) for the retrieval of personally important information; f) for the retrieval of information outlined by the teacher; g) for personal interests (like drawing, music, etc.)* (B37) (National student achievement test, 2008).

In accordance with the previous questions, the following areas of computer-aided activity were distinguished: playing games (*for playing computer games*); learning (*for learning*); communication (*for communication (e-mails, skype, chats, etc.);* creative leisure (*for personal interests (like drawing, music, etc.)*; retrieval of information important for the teacher (*for the retrieval of personally important information);* and retrieval of information important for the learner (*for the retrieval of personally important information; for the retrieval of information outlined by the teacher).*

All the answers of the respondents regarding the issues analysed in the current article were presented within the ordinal scale. Correlation analysis was applied to identify statistical correlation between the motivation of tenth formers to study Physics (*I like Physics*) and questions on ICT application and experimental activity in Physics.

The qualitative research was used to analyse motivation of tenth formers for learning Physics and its determinant factors. In accordance with the sociological theory of generations, these learners belong to Generation Z (born in 1995–2012). A semi-structured interview revealed the attitude of tenth formers towards the accomplishment of virtual and real laboratory work, as well as its role for motivation to study Physics. The learners were surveyed after the completion of two laboratory works that complied with the structured inquiry; and after completing two virtual laboratory works that also complied with the structured inquiry. The sample of the qualitative research was purposive modal instance sampling (Patton, 2002).

3.1.3 Results of the research

3.1.3.1 Factors determining motivation of Y generation learners for learning Physics

The analysis of 2008 data indicated statistically significant correlation between learners' replies about internal motivation (*I like Physics*) and performing Physics

Tab. 3.1.2: -Spearman's correlation coefficients of learners' experiential activity in the classes of Physics and their motivation for learning Physics (correlations have been calculated using the data base of the national research of 2008)

			Experiential activity		
	Motivation	**Physics grade**	**Performance labs in physics**	**Using different physical devices**	**Observation of experiments performed by the teacher**
Motivation (*I like Physics*)	1.000	0.182(**)	0.256(**)	0.249(**)	0.192(**)
Physics grade		1.000	0.051	0.062	0.042
Performance labs in physics			1.000	0.724(**)	0.570(**)
Using different physical devices				1.000	0.666(**)
Observation of experiments performed by the teacher					1.000

** Correlation is significant at the 0.01 level (2-tailed).

labs ($r = 0.256$; $p^{**} = 0.01$) as well as between motivation and use of different physical devices ($r = 0.249$; $p^{**} = 0.01$) (Tab. 3.1.2). However, a statistically insignificant correlation was established between the learners' replies regarding the motivation for learning Physics (*I like Physics*) and observation of experiments conducted by the teacher ($r = 0.192$; $p^{**} = 0.01$). This implies that experiments performed by the teacher do not carry statistical significance for the motivation for learning Physics of Y generation learners (Tab. 3.1.2). A conclusion could be drawn that learners are mainly motivated by their own experimental activity.

The data of national research (2004) has been analysed in order to disclose factors determining the motivation for learning physics of Y generation learners. The correlation coefficients between inquiry-based activity (planning and performing labs in Physics using different physical devices); computer related activity (playing computer games, e-communication with friends, and e-learning); motivation for learning Physics (I like Physics) and learning outcomes in Physics (semester grade for Physics) have been calculated (Tab. 3.1.3).

Statistically significant but weak correlations have been identified between the internal learners' motivation for learning Physics (I like Physics) and inquiry-based activities of planning and performing physics labs ($r = 0.238$; $p^{**} = 0.01$)

Tab. 3.1.3: Spearman's correlation coefficients of the motivation for learning Physics. inquiry-based activity in Physics and computer related activity of tenth formers (correlations have been calculated using the data base of the national research of 2004)

	Motivation		**Inquiry-based activity**		**Computer related activity**		
		Physics grade	**Planning. performing labs**	**Using physical devices**	**Games**	**Communication**	**Learning**
Motivation	1.000	0.212(**)	0.238(**)	0.205(**)	−0.178(**)	−0.012	−0.159(**)
Physics grade		1.000	0.072(**)	0.082(**)	0.108(**)	−0.115(**)	−0.103(**)
Planning performing labs			1.000	0.441(**)	−0.068(*)	−0.053(*)	−0.114(**)
Using physical devices				1.000	−0.059(*)	−0.048	−0.145(**)
Games					1.000	0.119(**)	0.195(**)
Communication						1.000	0.272(**)
Learning							1.000

** Correlation is significant at the 0.01 level (2-tailed).
* Correlation is significant at the 0.05 level (2-tailed).

as well as between motivation and using different physical devices (r = 0.205; p** = 0.01). Also a statistically significant but weak correlation has been estimated between the learners' positive motivation towards the subject of Physics (I like Physics) and the semester assessment (mark) for the subject (r = 0.212; p** = 0.01). As expected a strong correlation has been discovered between activity of planning physics labs and using physical devices (r = 0.441; p** = 0.01).

It is noteworthy that using the research data of 2008, statistically insignificant correlations was established between motivation for learning Physics (*I like Physics*) and semester assessment for Physics (Physics grade). This leads to an idea that the motivation for learning Physics of Y Generation learners does not statistically significantly correlate with the final learning outcomes in Physics. Therefore, learners with poorer learning outcomes in Physics can demonstrate positive motivation to learn the subject. A weak but statistically significant correlation (r = 0.212; p** = 0.01) was established while analysing the results of 2004 survey. Summarising the results of both surveys (2004 and 2008), a conclusion could be drawn that the statistical correlation between learning motivation and assessment in Physics was decreasing.

Summing up the obtained data, it could be concluded that information retrieval promoted the motivation for learning Physics of Y Generation learners. Learners searched for information in pursuance to answer encountered questions and seeking for new knowledge. This is the very essence of inquiry-based learning. Inquiry implies that the interaction with the environment is guided by a question; it is when we are in an indeterminate situation that we engage in inquiry in order to transform the problem encountered into something that we can grasp and act on (Petersson, Andersson, & Säljö, 2013).

3.1.3.2 Factors determining motivation of Z generation learners for learning Physics: Qualitative research results

A qualitative research was conducted in 2013 (a semi-structured interview), which involved 40 tenth formers. The tenth form was selected pursuing to compare the results of qualitative research with the data of previously conducted national research of tenth formers' achievements in Lithuania (in 2004 and 2008). Therefore, learners of the same age group were selected (16-year-olds), who represented a different generation: the learners of 2013 survey belonged to Generation Z, whereas the learners of 2004 and 2008 surveys represented Generation Y. The aim of the interview was to reveal the interest of Generation Z learners in Physics, as well as the effect of virtual and real laboratory experiments on their motivation.

The results of the qualitative research demonstrated that more than a half (59 %) of the informants of Generation Z (2013) liked Physics. Similarly, in

Tab. 3.1.4: Advantages of virtual laboratory work in Physics: statements and categories

Categories	Statements
Simple and clear technique of labs	'It is easier to perform Physics labs on the computer. you don't have to take care of the devices.' 'It is much easier to perform Physics labs on the computer. since all the information is provided in detail.' 'There is less mess.'
Fast accomplishment of the work	'Performance of Physics labs on the computer is easier and less time-consuming. you don't have to take care of the devices.' 'It is easier. simpler and faster.' 'I think that it is much faster to conduct laboratory work on the computer. and much simpler than in the classroom using special devices. Besides. you can use the devices only in the classroom. and the computer can be used at home.'
Individual speed of performing Physics labs	'I prefer to perform Physics labs using a computer because there is no need to hurry rather than doing the work with others in the classroom.'

the national survey of learner achievements (2004), 52 % of the respondents of Generation Y maintained liking Physics. Summarizing the obtained results, it is possible to state that about a half of research participants representing different generations (Y and Z) took interest in Physics. There was a nine-year gap between the two surveys. Consequently, the motivation of the target age group to study Physics experienced no considerable changes in a decade.

The attitude of informants towards real (performance of physics labs with real devices) and virtual (computer-aided) labs was explored. It was determined that 42 % of informants believed that laboratory work in Physics was better accomplished on a computer. It is noteworthy that this opinion was characteristic of the learners disliking Physics. Summarizing the answers provided by the informants about the advantages of performing Physics labs with a computer, the following categories could be distinguished: simple and clear technique of performing, fast accomplishment of the work, and individual speed of performance of Physics labs (Tab. 3.1.4).

The learners were asked how often they used computers and for what purposes (playing games, communication, learning, learning Physics, and creation). Having analysed the records of learners' replies it has appeared that the computer is used daily. All the informants (100 %) have mentioned using the computer for daily communication. Hence, leaners of Generation Z considered the computer as a significant means of communication. The informants' replies have showed that the computer is used for communication more frequently than for playing games. About half of learners (50 %) have stated that they play computer

Tab. 3.1.5: Advantages of laboratory work in Physics accomplished in the real settings: statements and categories

Categories	Statements
Hands–on dimension	'It is more interesting for me to touch. calculate and measure everything myself'. 'Real laboratory work is much better. While accomplishing it. you can touch. fiddle and try yourself. It is much more interesting'. 'I prefer doing everything with my own hands and observe physical laws in reality. or arouse them myself'. 'because I like touching and fiddling the materials myself. Working with real materials is more interesting'.
Sense of reality	'Doing something yourself is more interesting. you can examine everything'. 'to conduct experiments in reality is more interesting than doing them on the computer'. 'performing Physics labs using conventional devices is much more interesting: you can feel like a professor. It is better to observe different phenomena in reality'.
Application of teaching materials in practice	'Performing Physics labs in reality helps me to better understand theoretical materials and apply them in practice'. 'working with conventional devices. you can perform laboratory work yourself. learn the new information and apply the knowledge of Physics in an easier way'.

games every day. It has also appeared that the motivation to study Physics *(I like Physics)* is related to the interest in computer games. The learners who do not like Physics play computer games less frequently. None of these informants admitted playing computer games every day; whereas all the learners that liked Physics confessed playing computer games every day. There is a considerable number of surveys on the role of subject-specific computer games for the motivation for learning Physics. The role of games containing no specific content has not yet been thoroughly investigated and demands new research.

More than a half of the informants of the qualitative survey (58 %) stated that they liked Physics. It is noteworthy though that semester assessments for Physics were moderate ones. The same regularity can be observed as when accomplishing the secondary data analysis of the 2008 survey. A statistically insignificant correlation was established between the learners' interest in Physics and their semester assessments ($r = 0.182$; $p^{**} = 0.01$).

It was determined that learners who liked Physics preferred performing real labs in Physics. Having analysed the data of interview records, the most frequent statements of informants have been selected which are related to performing real Physics labs. In this regard, the following categories were distinguished: hands-on dimension, sense of reality, and application of teaching materials in practice (Tab. 3.1.5).

The analysis of informants' replies demonstrates that by indicating the advantages of real laboratory work, the learners most often signify the hands-on dimension: *I prefer doing it with my own hands*, *measure*, *fiddle*, and *touch*. It seems that the informants of Generation Z are bored with the traditional hand motoric employed in managing the mouse or managing a programme by touching. The informants of Generation Z preferred the activity, which would ensure more diverse hand motoric: touching, fiddling, and doing with hands. *Importance of 'hands-on' dimension* lies in the fact that 'hands-on' activities are central in an inquiry process and can be proposed virtually (Klahr, Triona, & Williams, 2007; Zacharia, Olympiou, & Papaevripidou, 2008). The informants' ideas showed that the accomplishment of laboratory work using real devices created the sense of reality (*When you conduct the laboratory work yourself, you can get involved into the process of its accomplishment and materials)*, and ensured the joy of discovery *(performing Physics labs using conventional devices is much more interesting: you can feel like a professor)*. According to the informants, laboratory work conducted in the real settings developed an ability to apply theoretical knowledge in practice.

3.1.4 Conclusions

Motivation for learning Physics, expressed by comprehensive school learners representing Generation Y, is conditioned by inquiry-based learning and experimental activity. Experiments conducted by the teacher have no statistically significant effect on the motivation for learning Physics. Learners are motivated mainly by their own performing experimental activity.

Motivation for learning Physics of Generation Y learners bears no statistically significant correlation with the use of the computer in afterschool activities: e-communication, computer games, e-learning, and creative leisure. However, the motivation for learning Physics of Generation Y learners is promoted by information retrieval from the virtual platform. Information retrieval aims at answering the encountered questions and solves subject-specific problems. This is an important stage of inquiry-based learning, which enhances learners' motivation to study Physics.

The qualitative research of tenth formers, learners representing Generation Z shows that more than a half of informants (59 %) express their interest in Physics. Virtual communication is more significant for Generation Z leaners, as they use the computer for e-communication every day. Moreover, virtual communication is more popular among learners of Generation Z than playing computer games.

The learners of Generation Z, who like Physics, prefer conducting real rather than virtual laboratory work in Physics. They maintain that performing laboratory work in reality creates conditions for 'lifelike' exploration by touching and fiddling the devices and materials (hands-on dimension), as well as permits experiencing the sense of reality and applying theory in practice.

The learners of Generation Z, who do not like Physics, prefer performing virtual laboratory work. They highlight the uncomplicated and clear technique, fast accomplishment as well as individual speed of conducting virtual laboratory work. Learners of Generation Z, who do not take particular interest in Physics, play computer games less frequently than the ones interested in the subject.

3.2 Digital and hands-on labs in Physics education at school: how do they engage?

By Palmira Pečiuliauskienė

The section deals with the role of real and digital Physics labs in the promotion of intrinsic motivation of lower secondary school students for learning Physics. The article examines how real and digital Physics labs meet basic psychological needs for autonomy, competency, and social relatedness of eighth-form learners.

In science education, supported by theories of constructivism there is a widely-held belief in the value of experimentation and laboratory work (Corter, Nickerson, Esche, Chassapis, Im, & Ma, 2007; Lang, 2012). The subject of Physics is an integral part of science. The problem of the attractiveness of the subject of Physics is very wide, it is analysed from different aspects: the individualisation of learning (Zacharia & Olympiou, 2010), collaborative learning (Nedic, Machotka, & Nafalski, 2003), formation of the concepts of Physics (Bajpai, 2013), formation of students scientific abilities (Etkina et al., 2010); formation of students motivation by digital and hands-on Physics labs (Lawlor et al., 2015; Peciuliauskiene, 2015).

Research conducted with 224 students from two large universities show no evidence that one of the treatments (virtual or hands-on) was more effective than the other in conveying the concepts of the labs to the students. There was no significant difference noted in any of the tests (Darrah et al., 2014).

According Lowe et al. (2013) there is lack of explicit consideration of students' attitudes towards practical experimentation of much of the broader literature. There is a lot of work that has focused on students' attitudes towards science generally, but only limited research about attitudes towards laboratories. Laboratory experimentation has generally been regarded as an important element of science education. Labs are suggested as one way to address the much discussed low knowledge levels and student interest in science (Shumov et al., 2013).

This situation necessitates for a deeper look into the problem of motivation for learning Physics - by the aspect of intrinsic motivation. The discussed situation highlights *the scientific problem* which is formulated as a question: What is the intrinsic motivation of the new generation learners for learning Physics and how is it determined by the experimental activity based on digital and hands-on technologies?

The object of the research is the intrinsic motivation of learners for learning Physics.

The aim of the research is to reveal the effects of digital and hands-on Physics labs on the intrinsic motivation of basic school learners for learning Physics.

The objectives of the research are as follows:

1. What is the impact of digital and hands-on Physics labs on the intrinsic motivation of basic school learners?
2. What psychological learner needs determined by Self-Determination Theory (SDT) are assured by digital and hands-on Physics labs?
3. How is the motivation of new generation learners for learning Physics related to basic psychological needs determined by SDT, i.e. a need for autonomy, a need for competency and a need for social interaction?

3.2.1 Method of research

The research methodology is based on constructivist theory of education, which acknowledges structured inquiry as an efficient educational technology promoting a positive attitude towards the subjects of natural sciences and helping to apply the acquired knowledge in different situations, developing higher-level thinking abilities as well as promoting active learning processes that are based on knowledge and experience. Moreover, realist education philosophy stating that the reality of natural sciences is objective and cognisable is considered.

Intrinsic Motivation Inventory (IMI), a valid and reliable instrument, was used to explore intrinsic motivation. McAuley, Duncan, and Tammen (1987) carried out a study to examine the validity of the IMI and found strong support for its validity. IMI is a multidimensional measurement device intended to assess participants' subjective experience related to a target activity in laboratory experiments (Ryan, 1982). This instrument allows assessing intrinsic motivation and self-regulation of learners. The first subscale (*interest/enjoyment*) is the main subscale of IMI and assesses the intrinsic motivation of learners. On the basis of this subscale, overall questionnaire is called IMI. According to the SDT, the second subscale *feeling of competency*, the fifth subscale *perceived choice* and the seventh subscale *interpersonal interactions* are important for intrinsic motivation. The results of subscales are represented by the interval scale, which ranges from 1 to 100 points.

The research sample is reliable and representative (probability cluster sample). The sample included eighth-form learners of Lithuania. The research clusters were the largest cities of Lithuania. Classes were selected on the basis of probability cluster sample and all learners of a selected class were tested.

The total population of eighth-form students in 2014–2015 was 25,000 (Education Management Information System – EMIS). The confidence interval being 5 %, confidentiality level is 95 %. Hence, the research sample should have included 379 respondents. The research sample was reliable as it involved 385 students. A reliable sample of 385 respondents gives a possibility for data generalization. Eighth-form students were tested using IMI questionnaire after digital and after real Physics lab.

The learners accomplished a Physics lab using digital laboratory software Xplorer GLX. It is a tool of storage, presentation, and analysis of the data of experimental measurements that operates with different sensors. The learners also accomplished a Physics lab using real devices.

The learners were working in groups: on average three persons per group. In terms of inquiry-based levels, the lab conformed to the second level (structured inquiry). Before the accomplishment of the lab, the learners were introduced to the aim and procedure of the work, but they were not familiar with the result. After the accomplishment of the lab, the learners filled in IMI questionnaire that was meant to determine intrinsic learning motivation and its determinant factors.

3.2.2 Results

3.2.2.1 Results on digital Physics labs

The study focused on the promotion of learners' intrinsic motivation to study Physics by digital Physics labs. We used IMI questionnaire which consisted of seven groups of questions – subscales (Tab. 3.2.1). From the viewpoint of SDT, the following subscales are important: feeling of competency, perceived choice, and interpersonal interactions. There are two scales of IMI that are related to personal effort and felt pressure and tension (Tab. 3.2.1) and one subscale related to value/usefulness.

A 100 point interval measurement scale was used to explore the intrinsic motivation of learners; therefore, parametrical data were obtained. On this basis, the mean, median, and standard deviation were calculated for every subscale

Tab. 3.2.1: Key tendencies of learners' intrinsic motivation and its determinant factors

	1. Interest/ enjoyment	2. Feeling of competency	3. Effort	4. Felt pressure and tension	5. Perceived choice (autonomy)	6. Value/ usefulness	7. Interpersonal interactions
Mean	59.74	62.73	64.91	52.48	59.71	76.67	52.00
Median	60.00	63.00	64.00	52.00	60.00	80.00	55.00
Std. Deviation	11.590	14.272	13.941	12.814	13.032	23.392	19.632
Skewness	0.074	−0.443	−0.370	0.217	0.162	−0.905	−1.064
Std. Error of Skewness	0.295	0.295	0.295	0.295	0.295	0.295	0.295
Minimum	37	30	24	20	31	20	0
Maximum	80	90	100	92	91	100	90

(interest/enjoyment, feeling of competency, effort, felt pressure and tension, perceived choice, value/usefulness, interpersonal interactions), as well as skewness coefficients of every subscale were drawn (Tab. 3.2.1).

The first subscale (interest/enjoyment) is the most significant one in IMI (Tab. 3.2.1) as it reveals learners' intrinsic motivation for learning Physics. Its mean is 59.74 points out of the total of 100 points. The research data demonstrate that the data of the subscale is distributed according to the normal distribution and is marked by positive skewness (interest/enjoyment). The mean of the first subscale (interest/enjoyment) is closest to the mean of the fifth subscale (perceived choice or autonomy) (59.71). Distribution of the fifth subscale is also marked by a weak positive skewness and it approximates the normal distribution.

It should be stated that the assessment of the satisfaction of the basic psychological needs (autonomy, competency, and interpersonal interaction) differs while conducting digital labs in Physics. Digital Physics labs mostly satisfy learners' need for the competency feeling: the mean of the second subscale is 62.73. Therefore, digital Physics labs provide preconditions for a learner to feel competent: "I think I am pretty good at this activity"; "I think I did pretty well at this activity, compared to other learners". It appeared that a need for interpersonal interaction is satisfied least of all by digital Physics labs. The mean of the seventh subscale equals to 52.00. The discrepancy between the second and seventh subscales equals to 10.73 (Tab. 3.2.1).

The research data reveal (Tab. 3.2.1) that while conducting digital Physics labs learners do not experience extensive emotional pressure and tension (the mean is 52.48). Nevertheless, accomplishment of digital Physics labs demands some effort (the mean is 64.91). According to IMI questionnaire, it is impossible to determine what demands more effort: the content of Physics lab or its digital means. Therefore, it is impossible to draw a conclusion that learners of the new generation consider digital Physics labs as demanding effort.

The study also attempted at exploring whether the means of the subscales differed statistically significantly from the point of view of SDT (Tab. 3.2.2). ANOVA (Analysis of Variance) data block was used to identify statistical significance of the differences in means. This statistical criterion is applied for more than two dependent samples when the data are parametrical.

Sphericity Assumed ($p = 0.000$) showed means that differed statistically significantly. Significant statistical differences of the means of the second (feeling of competency), fifth (perceived choice or autonomy), and seventh (interpersonal interactions) subscales were explored (Tab. 3.2.2).

The results of ANOVA Bonferoni test in the table of Pairwise Comparisons demonstrated statistically significant differences of the means of the seventh

Tab. 3.2.2: Results of ANOVA Bonferoni data block test. Pairwise Comparisons between main factor of Self-Determination theory: 2 – feeling of competency; 5 – perceived choice or autonomy; 7 – interpersonal interactions

(I) factor1	(J) factor1	Mean Difference (I–J)	Std. Error	Sig.[a]	95 % Confidence Interval for Difference[a] Lower Bound	Upper Bound
2	5	3.015	2.204	0.528	−2.400	8.431
	7	10.727*	2.718	0.001	4.047	17.408
5	2	−3.015	2.204	0.528	−8.431	2.400
	7	7.712*	2.611	0.013	1.297	14.128
7	2	−10.727*	2.718	0.001	−17.408	−4.047
	5	−7.712*	2.611	0.013	−14.128	−1.297

Based on estimated marginal means
[a] Adjustment for multiple comparisons: Bonferroni.
* The mean difference is significant at the ,05 level.

(interpersonal interactions) subscale from the second (feeling of competency) and fifth (perceived choice or autonomy) subscales (Tab. 3.2.2). The statistically significant difference occurred due to the fact that the mean of the seventh (interpersonal interactions) subscale was lower than those of the second and fifth subscales (Tab. 3.2.1). It was determined that the need of the learners of new generation for social interaction was less important in experimental activity in comparison to the feeling of competency ($p = 0.001$) or the feeling of autonomy of activity ($p = 0.013$) (Tab. 3.2.2).

Correlation between the data of different subscales was also explored (Tab. 3.2.3). The strongest statistically significant correlation was determined between learners' intrinsic motivation for learning Physics and the feeling of competency (the first and second subscales) ($r = 0.462^{**}$, $p = 0.01$). Hence, digital labs in Physics gave learners' the feeling of competency and promoted positive motivation for learning Physics.

A statistically significant correlation was determined between learners' intrinsic motivation for learning Physics and the feeling of autonomy while accomplishing a Physics lab (the first and fifth subscales) ($r = 0.235^{*}$, $p = 0.01$). This correlation is significant in assessing the reliability of the questionnaire for motivation. In fact, to be confident in one's assessment of intrinsic motivation, one needs to find that the free-choice behaviour and the self-reports of interest/enjoyment are significantly correlated (Ryan, Koestner, & Deci, 1991). A statistically insignificant correlation was determined between intrinsic motivation for learning Physics (the first subscale) and interpersonal interaction (the seventh subscale) ($r = 0.187$, $p = 0.05$).

Correlational analysis demonstrated (Tab. 3.2.3) that digital labs in Physics created the feelings of activity value and usefulness and established strong and statistically significant correlation with intrinsic motivation for learning Physics ($r = 0.445^{**}$, $p = 0.01$). The data on the value/usefulness of experimental activity was marked by strong and statistically significant correlations with the data of the subscale of feeling of competency ($r = 0.377^{**}$, $p = 0.01$) and with the one of effort ($r = 0.339^{**}$, $p = 0.01$). Value/usefulness of experimental activity enhanced the feeling of competency of the learners of Generation Z; however, it demanded more effort. Three statistically significant correlations that were determined in the subscale of value/usefulness suggested that the learners of Generation Z were characterized by a pragmatic approach to experimental activity.

According to Ryan, Koestner, & Deci (1991), when participants were ego involved, they engaged in pressured persistence during a free-choice period and this behaviour did not correlate with the self-reports of interest/enjoyment. Conclusions of Ryan, Koestner, and Deci (1991) were drawn with reference to the studies of intrinsic motivation of preceding generations (Generation X or Generation Y). The research of new generation learners' intrinsic motivation for learning Physics that was based on the promotion of motivation for learning demonstrated (Tab. 3.2.3) that the learners were ego involved because the data of the fourth subscale (felt pressure and tension) did not statistically significantly correlate with the data of the first subscale (interest/enjoyment).

3.2.2.2 Results on hands-on Physics labs

The investigation focused on the analysis how real Physics labs determine the students' intrinsic motivation for learning, feeling of competency, feeling of autonomy, and students' interpersonal communication while accomplishing a real Physics lab. As in the case of digital physics laboratory work a 100-point interval measurement scale was used to explore the intrinsic motivation of learners; therefore, parametrical data were obtained. On this basis, the mean, median, and standard deviation were calculated for every subscale (interest/enjoyment, feeling of competency, effort, felt pressure and tension, perceived choice, value/usefulness, interpersonal interactions), as well as skewness coefficients of every subscale were drawn (Tab. 3.2.4).

The first subscale (Interest/enjoyment) is the most significant in IMI questionnaire, since it reveals the students' intrinsic motivation for learning Physics.

Tab. 3.2.3: Promotion of the learners' intrinsic motivation for learning Physics through digital experimental activity: Pearson correlation coefficients between IMI questionnaire subscales

	Interest/ enjoyment	**Feeling of competency**	**Effort**	**Felt pressure and tension**	**Perceived Choice**	**Value/ usefulness**	**Interpersonal interactions**
Interest/enjoyment	1	0.462**	–0.195	0.019	0.235*	0.445**	0.187
Feeling of competency		1	0.373**	0.072	0.143	0.377**	0.181
Effort			1	0.123	0.422**	0.339**	0.288*
Felt pressure and tension				1	0.264*	–0.092	0.280*
Perceived Choice					1	–0.025	0.206
Value/usefulness						1	–0.125
Interpersonal interactions							1

** Correlation is significant at the 0.01 level (2-tailed).
* Correlation is significant at the 0.05 level (2-tailed).

Its value was 62.47 points out of 100. The research data revealed that the subscale data followed normal distribution and were marked by positive asymmetry.

The highest point was attributed to the sixth subscale – value/usefulness – 75.61 (Tab. 3.2.4). Hence, the new generation students understood the significance of real Physics labs and recognised their practical value. The fourth subscale – felt pressure and tension – received the lowest value, which was 53.27 (Tab. 3.2.4). It means that Physics labs created a positive pedagogical-physiological environment.

The meaning of the first subscale (interest/enjoyment) was closely related to the meaning of the second subscale (feeling of competency) – 63.29. Hence, real Physics labs mostly satisfied the students' feeling of competency. Real Physics labs gave an opportunity for the students to feel competent. The comparison of the students' intrinsic motivation while performing real (Tab. 3.2.4) and digital (Tab. 3.2.1) Physics labs demonstrated that real Physics labs (62.47) were recognised as more interesting than digital (59.74) by the new generation students.

It is noteworthy that the main psychological needs that are essential for need satisfaction (perceived choice/autonomy, feeling of competency, and social interaction) differed when performing real Physics labs. It appeared that the need for social interaction during the accomplishment of a real Physics lab was least satisfied, since the value of the seventh subscale was 54.74 (Tab. 3.2.4). However, the need for social interaction while accomplishing a real Physics lab was satisfied better than while accomplishing a digital Physics lab – 52.00 (Tab. 3.2.1).

Having compared the feeling of competency while accomplishing real and virtual Physics labs, insignificant differences were determined: performing digital Physics labs – 62.73 (Tab. 3.2.1), whereas performing real Physics labs – 63.29 (Tab. 3.2.4). The students' feeling of autonomy while performing real (59.71) and digital Physics labs (59.12) differed merely in the decimal significances after the comma (Tab. 3.2.1; Tab. 3.2.4).

The study also attempted at exploring whether the means of the subscales differed statistically significantly from the point of view of SDT (Tab. 3.2.5). ANOVA data block was used to identify the statistical significance of the differences in means. It was mentioned that this statistical criterion is applied for more than two dependent samples when the data are parametrical.

Bearing in mind the research data, ANOVA is applicable, since the distribution of all subscales is marked by negative asymmetry. Therefore, all of them approximate to the normal distribution (Tab. 3.2.4).

The results of Sphericity Assumed showed means that differed statistically significantly ($p = 0.000$). Significant statistical differences of the means of the

Tab. 3.2.4: Key tendencies of learners' intrinsic motivation at real physics labs and its determinant factors

	1. Interest/ enjoyment	**2. Feeling of competency**	**3. Effort**	**4. Felt pressure and tension**	**5. Perceived choice (autonomy)**	**6. Value/ usefulness**	**7. Relatedness**
Mean	62.47	63.29	63. 70	53.27	59.12	75.61	54.74
Median	66.00	63.00	64.00	52.00	60.00	80.00	55.00
Std. Deviation	12.563	13.959	11.914	11.702	13.605	20.372	11.283
Skewness	–0.997	–0.404	–0.169	–0.401	–0.131	–0.687	–0.715
Std. Error of Skewness	0.295	0.295	0.295	0.295	0.295	0.295	0.295
Minimum	37	30	24	20	31	20	0
Maximum	80	90	100	92	91	100	90

Tab. 3.2.5: Results of ANOVA Bonferoni data block test at real Physics labs. Pairwise comparisons between the main factors of Self-Determination Theory: 2 – Feeling of competency; 5 – Perceived choice or autonomy; 7 – Relatedness

(I) factor1	(J) factor2	Mean Difference (I–J)	Std. Error	Sig.[a]	95 % Confidence Interval for Difference[a]	
					Lower Bound	Upper Bound
2	5	4.167	2.372	0.251	−1.663	9.997
	7	8.545*	2.004	0.000	3.621	13.470
5	2	−4.167	2.372	0.251	−9.997	1.663
	7	4.379	2.300	0.840	−1.274	10.031
7	2	−8.545*	2.004	0.000	−13.470	−3.621
	5	−4.379	2.300	0.184	−10.031	1.274

Based on estimated marginal means
[a] Adjustment for multiple comparisons: Bonferroni.
* The mean difference is significant at the ,05 level.

second (feeling of competency), fifth (perceived choice or autonomy) and seventh (relatedness) subscales were explored (Tab. 3.2.5).

The results of ANOVA data block test revealed statistically significant differences between the second subscale (feeling of competency) and the seventh subscale (relatedness). A statistically significant difference occurred due to the fact that the estimated value of the seventh subscale (relatedness) was lower than the one of the second subscale (feeling of competency). Hence, the need for social interaction while performing real Physics labs was less significant for new generation (Z) students than the feeling of competency ($p = 0.000$) (Tab. 3.2.5). In all other compared cases, statistically insignificant differences were determined (Tab. 3.2.5).

Correlation between real Physics labs data of different subscales was explored (Tab. 3.2.6). According to SDT, humans actively seek for the actualisation of their potentialities fulfilling their basic psychological needs: needs for autonomy, competency, and social relatedness. In this case, it was important to determine how the data of the first (interest/enjoyment) subscale correlated with the data of the second (feeling of competency), fifth (perceived choice) and seventh (interpersonal interactions) subscales.

The statistically strongest correlation was detected between intrinsic motivation (interest/enjoyment) of students and feeling of value/usefulness of real Physics labs ($r = 0.492^{**}$, $p = 0.01$). It means that real Physics labs are useful for new generation students and engage them for learning Physics.

Tab. 3.2.6: Promotion of the learners' intrinsic motivation for learning Physics through real experimental activity: Pearson correlation coefficients between IMI questionnaire subscales

	1. Interest/ enjoyment	**2. Feeling of competency**	**3. Effort**	**4. Felt pressure and tension**	**5. Perceived choice (autonomy)**	**6. Value/ usefulness**	**7. Relatedness**
Interest/ enjoyment	1	0.456**	–0.037	0.013	0.041	0.492**	0.236
Feeling of competency		1	0.188	0.069	0.020	0.315**	0.153
Effort			1	–0.057	–0.063	0.119	0.091
Felt pressure and tension				1	0.427**	–0.017	0.090
Perceived choice (autonomy)					1	–0.175	0.149
Value/usefulness						1	0.226
Relatedness							1

** Correlation is significant at the 0.01 level (2-tailed).
* Correlation is significant at the 0.05 level (2-tailed).

A statistically significant correlation was determined between the students' intrinsic motivation for learning Physics and the feeling of competency while performing a real Physics lab (the first and second subscales) ($r = 0.456^{**}$, $p = 0.01$). Hence, similarly to digital Physics labs (Tab. 3.2.3), real Physics experiments (Tab. 3.2.6) stimulated the students' feeling of competency and positive motivation for learning Physics. The correlation between their intrinsic motivation for learning Physics and the feeling of value/usefulness was stronger than the correlation between motivation for learning Physics and the feeling of competency while performing a lab.

The data of value/usefulness in real experimental activity were characterised by a strong and statistically significant correlation with the data of the subscale of the feeling of competency ($r = 0.315$, $p = 0.01$). The two statistically significant correlations that were detected on the subscale of value/usefulness prove that the new generation students have a pragmatic approach to experimental activity.

3.2.3 Conclusions

Digital labs in Physics are most often accomplished at school. Their impact on the intrinsic motivation of new generation (Z) learners, or the so-called learners of technology generation, is significant; however, it is not essential. According to IMI questionnaire, the mean of learners' intrinsic motivation for learning Physics is 59.74 points out of the total of 100 points. Basic school learners are convinced about the value, practical applicability, and benefit for the future of digital labs in Physics. The mean of the subscale of value/usefulness is 76.67 out of the total of 100 and is higher than that of other subscales (of IMI). Awareness of the value of experimental activity is an important factor of a new generation learner's intrinsic motivation for learning Physics.

The peculiarities of the intrinsic motivation of new generation learners for experimental activity are revealed on the basis of SDT. According to this theory, intrinsic motivation for learning occurs when basic psychological needs (for autonomy, competency, and social relatedness) are satisfied. Digital labs in Physics provide highest satisfaction of new generation learners' need for competency, and the lowest satisfaction for the need of interpersonal interactions (the difference is statistically significant: $p = 0.001$). Moreover, digital labs in Physics provide better satisfaction of the need for autonomy as compared to the need of interpersonal interactions (the difference is statistically significant: $p = 0.0013$).

Correlational analysis of the main components of SDT reveals that the strongest and statistically significant correlation is manifested between learners' intrinsic motivation for learning Physics and their feeling of competency (r = 0.462**, when p = 0.01), as well as between intrinsic motivation for learning Physics and the feeling of autonomy while accomplishing a digital lab (the first and fifth subscales) (r = 0.235*, when p = 0.01). The correlation between learners' intrinsic motivation for learning Physics and their feeling of competency is stronger than the correlation between motivation to learn Physics and the feeling of autonomy while accomplishing a digital lab.

While performing real Physics labs, the basic psychological needs (autonomy, feeling of competency, and interpersonal communication) that are essential for students' intrinsic motivation for learning are satisfied differently. Real Physics labs tend to mostly satisfy the feelings of autonomy and competency of the new generation students, yet they demand a lot of effort for their accomplishment. The correlation analysis of the main components of SDT demonstrated that while accomplishing real Physics labs, the most prominent and statistically significant correlation was detected between the students' intrinsic motivation for learning Physics and its value/applicability; as well as between the intrinsic motivation for learning and the feeling of competency. The new generation students consider real Physics labs as valuable, since they create preconditions for the students' to feel competent.

The statistical correlations between IBL components and the abilities of information exchange depend on the form of experimental activity (real or virtual). Combining the real and virtual forms in the educational practice of Physics, different IBL components can be stipulated: motivation, feeling of competency, and the feeling of activity autonomy.

3.3 Formation of intrinsic motivation at different levels of inquiry

By Palmira Pečiuliauskienė

The section examines the role of Physics labs in the promotion of intrinsic motivation of lower secondary school students for learning Physics. The article focuses on the second level of inquiry-based learning referred to as structured inquiry and on the third level referred to as guided inquiry. Students' intrinsic motivation is analysed on the basis of SDT.

3.3.1 Introduction

Experts have long recommended implementing more laboratory activities in school science classes in order to engage students in science and learning (Hofstein & Lunetta, 2004). Experimental activity labs are suggested as one of the ways to promote students' interest in science (Bizzo, et al., 2002). According to Shumow et al. (2013), there has been surprisingly little recent research on laboratory lessons in school science classes in terms of the perceived value of laboratory activities.

Inquiry-based Science Education (IBSE) is a framework for conceptualising the priorities and values of laboratory activities. Firstly, inquiry-based learning offers opportunities for learners to achieve a better understanding of science phenomena. Within an inquiry context, students can develop an understanding of science concepts and ideas related to the phenomena experienced in everyday life. Employing inquiry strategies, *"teachers use student-centred instruction that is less step-by-step and teacher-directed than traditional instruction; students answer scientific research questions by analysing data (e.g., using guided or facilitated inquiry activities, laboratory inquiries)"* (Schroeder et al., 2007, p. 1446). Inquiry involves a degree of autonomy, or responsibility for learning, given to learners or teachers in progressing from a problem situation towards an answer, a solution or a higher level of knowledge. The common goal for science educators is to engage students in inquiry; however, many factors complicate the completion of such a task (Buck, Bretz, & Towns, 2008).

Many hands-on activities (constructing models of cells or atoms, or constructing a model of the solar system) performed in science classrooms do not correspond to the inquiry strategy as they do not involve a research question or data analysis. There is a lack of agreement about what constitutes an inquiry-based approach (Buck, Bretz, & Towns, 2008). It is important to realise that not all inquiry activities are equal. The concept of different levels of inquiry was first described by Schwab (1962). Later, Herron identified three levels of openness for inquiry in science activities (1971). About twenty years later, Rezba, Auldridge, and Rhea (1999) described a four-level model of inquiry instruction based on the work by Schwab and Herron. *National Science Education Standards* (NRC, 2000), as well as Brown et al. (2006) present inquiry as a continuum moving from more to less guidance.

"We analysed 22 laboratory manuals and nearly 400 experiments leading to the articulation of more specific levels of inquiry and more detailed characteristics. The characteristics of the rubric originated from two sources, the terminology used in laboratory manuals to organise components of a lab and the key elements in a laboratory activity where students might become independently engaged" (Buck, Bretz, & Towns, 2008, p. 53). Four levels of inquiry were described: confirmation inquiry, structured inquiry, guided inquiry, and open inquiry. As we go deeper into the levels, students learn more and become more engaged and self-confident. At the level of confirmation inquiry, students know the problem, ways and means of the research, as well as answers. Students confirm a principle through a prescribed activity when the results are known in advance. At the level of structured inquiry, students know the problem, ways and means of the research but they do not know the answer of the research. Students investigate a teacher-presented question through a prescribed procedure. At the level of Guided Inquiry, students know only the problem of the research. They investigate a teacher-presented question using student designed/selected procedures. At the level of Open Inquiry, the problem, procedures/design, analysis, communication, and conclusions are for a student to design.

In literature reviews, scholars have indicated the need to study motivational and affective dimensions of students' science experiences in general (Krapp & Prenzel, 2011) and in particular during lab experiences (Hofstein & Lunetta, 2004). "Each of those reviewers observed that there is a widespread belief and supporting evidence that affective and motivational experiences are essential to student learning, yet those subjective experiences have been under-studied in science education studies, especially those pertaining to lab activities" (Shumow et al., 2013, p. 233). According to Shumow et al. (2013), multiple perspectives were valuable for understanding laboratory activity as it unfolded in the classroom.

Two perspectives for understanding Physics laboratory activity were chosen. The scientific problem of the research is formulated as a question: how do structured and guided Physics labs engage new generation students?

The object of the research is intrinsic motivation of secondary school learners for learning Physics.

The aim of the research is to reveal the impact of structured and guided inquiry on intrinsic motivation of secondary school learners for learning science.

The objectives of the research are as follows:

1. How do structured and guided inquiry Physics labs engage school students in science?
2. How are the engagement variables related to *autonomy, competency, and social relatedness* of structured and guided Physics labs activity?

3.3.2 Methodology and method of research

The research methodology is based on the constructivist theory of education, which acknowledges structured inquiry as an efficient educational technology promoting a positive attitude towards the subjects of natural sciences and helping to apply the acquired knowledge in different situations, developing higher-level thinking abilities, as well as promoting active learning processes that are based on knowledge and experience. Moreover, realist education philosophy stating that the reality of natural sciences is objective and cognisable is considered.

The instrument of quantitative research. IMI, was used to explore intrinsic motivation. The results of subscales are represented by the interval scale, which ranges from 1 to 100 points. More information about IMI questionnaire is given in the Section 2.2: *Research methods and instruments of motivation for learning.*

The sample and sampling of quantitative research. The research sample is reliable and representative (probability cluster sample). The sample included eighth-form learners of Lithuania. The research clusters were the largest cities of Lithuania. The classes were selected on the basis of probability cluster sample and all learners of the selected class were tested.

The research sample was reliable. It involved 388 learners. The total population in 2014–2015 was 25,000 eighth-form students (EMIS). The 5 % confidence interval was chosen. Confidentiality level was 95 %. Hence, the research sample should have included 379 respondents. Eighth-form students were tested using IMI questionnaire after Structured and after Guided Physics labs. A reliable sample of 379 respondents gives a possibility for data generalization.

3.3.3 Results

3.3.3.1 Intrinsic motivation of school students for learning Physics at structured inquiry Physics labs

Seven subscales of IMI questionnaire were used in the research on structured Physics labs: interest/enjoyment' perceived competency' effort' value/usefulness' felt pressure and tension' perceived choice (or autonomy of activity), and relatedness. As it was mentioned earlier, the first subscale (*interest/enjoyment*) is the main subscale of IMI and assesses the intrinsic motivation of learners.

The parameters of descriptive statistics were estimated in accordance with the following subscales: the mean, standard deviation, and skewness (Tab. 3.3.1). The research data were parametric; the interval scale was employed.

The research data demonstrated that structured Physics labs evoked the students' intrinsic motivation ($\overline{x}$= 66.48 ± 12.92). The data of this subscale approximated the normal distribution and was marked by the negative asymmetry (Tab. 3.3.1). The highest rate was ascribed to the sixth subscale *value/usefulness* ($\overline{x}$= 73.09 ± 23.21) (Tab. 3.3.1). Consequently, it is obvious that the value/usefulness of structured Physics labs motivated new generation learners. The fourth subscale – *felt pressure and tension* has received the lowest rate ($\overline{x}$= 55.37 ± 12.63) (Tab. 3.3.1). Hence, Physics labs of the structured level created a favourable learning environment in terms of the emotional aspect.

The differences of the subscale means (Tab. 3.3.1) were evaluated statistically. In the case of dependent samples, ANOVA data block test was applied. The data

Tab. 3.3.1: Key tendencies of students' intrinsic motivation at real Physics labs at the level of structured inquiry and its determinant factors

	Interest/ enjoyment	**Feeling of competency**	**Effort**	**Felt pressure/ tension**	**Perceived choice (autonomy)**	**Value/ usefulness**	**Relatedness**
Mean	66.48	63.04	62.22	55.37	57.59	73.09	57.81
Std. Deviation	12.929	17.07	14.02	12.634	16.227	23.217	12.981
Skewness	−0.032	−0.124	0.113	0.168	0.016	−0.689	−0.214

Tab. 3.3.2: Verification of the normality standard of the research data: Kolmogorov-Smirnov testing coefficients in case of Physics labs at the structured inquiry level

	Interest/ enjoyment	**Feeling of competency**	**Effort**	**Felt pressure/ tension**	**Perceived choice (autonomy)**	**Value/ usefulness**	**Relatedness**
Kolmogorov-Smirnov Z	1.083	0.781	1.862	1.078	1.121	1.262	0.898
Asymp. Sig. (2-tailed)	0.192	0.576	0.052	0.195	0.162	0.083	0.396

obtained on the Physics labs at the structured inquiry level using IMI questionnaire complied with the normality standard of the Kolmogorov-Smirnov test ($p > 0.05$) (Tab. 3.3.2).

ANOVA data block test revealed a statistically significant difference between the means of the first subscale *interest/enjoyment* and the fourth subscale *felt pressure and tension;* as well as between the means of the fifth subscale *perceived choice* and the seventh subscale *relatedness* (Tab. 3.3.3).The first subscale estimates students' intrinsic motivation for learning while performing Physics labs at the level of structured enquiry. Consequently, students' intrinsic motivation is more expressed than the perceived tension, the feeling of autonomy, and the feeling of social interaction in a group.

The data analysis revealed a statistically insignificant difference between the data of the first subscale *interest/enjoyment* and the third subscale *effort* (Tab. 3.3.3). Moreover, a statistically insignificant difference was determined between the first (*interest/enjoyment*) and the sixth subscale (*value/usefulness*) ($p > 0.05$). Consequently, the students' feeling of enjoyment while performing Physics labs according to the structured inquiry statistically differs neither from the feeling of the allotted effort while performing a lab nor the feeling of perceived choice (Tab. 3.3.3).

Tab. 3.3.3: Results of ANOVA Bonferoni data block test at structured level Physics labs. Pairwise comparisons between main factor of Self-Determination Theory: 2 – feeling of competency; 5 – perceived choice or autonomy; 7 – relatedness

(I) factor1	(J) factor1	Mean Difference (I–J)	Std. Error	Sig.[a]	95 % Confidence Interval for Difference[a]	
					Lower Bound	Upper Bound
1. Interest/ enjoyment	2. **Feeling of competency**	3.436	1.511	0.527	−1.275	8.146
	3. Effort	4.257	1.514	0.125	−.463	8.978
	4. Felt pressure and tension	11.109*	1.621	0.000	6.056	16.162
	5. **Perceived choice (autonomy)**	8.881*	1.858	0.000	3.088	14.674
	6. Value/ usefulness	−6.614	2.210	0.073	−13.504	0.277
	7. **Relatedness**	8.663*	1.723	0.000	3.290	14.036

Tab. 3.3.4: Promotion of the students' intrinsic motivation for learning Physics through structured inquiry Physics labs: Pearson correlation coefficients between IMI questionnaire subscales

	Interest/ enjoyment	Feeling of competency	Effort	Felt pressure and tension	Perceived choice (autonomy)	Value/ usefulness	Relatedness
Interest/ enjoyment	1	0.516**	0.365**	0.188	0.195	0.354**	0.106
Feeling of competency		1	0.367**	0.027	0.222*	0.396**	0.052
Effort			1	0.191	0.385**	0.255*	0.252*
Felt pressure and tension				1	0.288**	0.082	0.413**
Perceived choice (autonomy)					1	0.391**	0.213*
Value/ usefulness						1	0.096
Relatedness							1

** Correlation is significant at the 0.01 level (2-tailed).
* Correlation is significant at the 0.05 level (2-tailed).

Tab. 3.3.5: Key tendencies of the students' intrinsic motivation at real Physics labs at guided inquiry level and its determinant factors

	Interest/ enjoyment	Feeling of competency	Effort	Felt pressure/ tension	Perceived choice (autonomy)	Value/ usefulness	Relatedness
Mean	63.8	57.31	61.7	52.86	59.74	75.66	54.57
Std. Deviation	12.513	16.721	10.149	12.542	14.246	23.341	11.234
Skewness	−0.707	−0.347	0.138	−0.278	−0.064	−0.618	0.075

The correlation between the data of different subscales at structured inquiry physics labs was explored (Tab. 3.3.4). The strongest statistically significant correlation was determined between the students' intrinsic motivation for learning Physics and the feeling of competency (the first and second subscales) ($r = 0.516^{**}$, $p = 0.01$). The same tendency was observed at guided inquiry Physics labs. Hence, structured inquiry Physics labs also gave students the feeling of competency and promoted positive motivation for learning Physics.

The correlations between the first subscale *interest/enjoyment* and *perceived choice/autonomy*, as well as *interest/enjoyment* and *relatedness* were statistically insignificant (Tab. 3.3.4). It means that at structured inquiry Physics labs intrinsic motivation for learning Physics promotes the feeling of competency. The feeling of autonomy (*perceived choice/autonomy*) and interpersonal communication in students' group labs does not correlate with intrinsic motivation for learning Physics.

3.3.3.2 *Intrinsic motivation of school students for learning Physics at guided inquiry Physics labs*

The influence of Physics labs performed at the level of guided inquiry on the motivation for learning Physics was explored. Trying to solve this issue, IMI questionnaire consisting of seven subscales was employed (Tab. 3.3.5).

The first subscale *interest/enjoyment* discloses intrinsic motivation of school students at guided Physics labs ($\overline{x} = 63.80 \pm 12.51$). The research data demonstrated that the subscale data were distributed in accordance with the normal distribution and were not characterised by a high asymmetry (Tab. 3.3.5).

The sixth subscale *value/usefulness* received the highest evaluation ($\overline{x} = 75.66 \pm 23.34$) (Tab. 3.3.5). Hence, guided level Physics labs motivate new generation students. Students consider labs accomplished at the guided inquiry level as valuable, and foresee their practical significance. The lowest point was attributed

Tab. 3.3.6: Verification of the normality of research data in Physics labs at guided inquiry level: Kolmogorov-Smirnov coefficients

	Interest/ enjoyment	Feeling of competency	Effort	Felt pressure / tension	Perceived choice (autonomy)	Value/ usefulness	Relatedness
Kolmogorov-Smirnov Z	1.464	0.895	1.101	1.225	0.681	1.788	1.488
Asymp. Sig. (2-tailed)	0.057	0.400	0.177	0.099	0.742	0.063	0.074

[a] Test distribution is Normal.
[b] Calculated from data.

to the fourth subscale *felt pressure and tension* ($\overline{x} = 52.86 \pm 12.54$) (Tab. 3.3.5). It means that guided inquiry Physics labs create a favourable emotional learning environment, in which students feel no tension and pressure.

Seeking to statistically compare the means of different subscale ranks, ANOVA data block test was applied as the research samples were independent. ANOVA data block test can be applied when research data comply with the normality standard. In our research, the data complied with the normality standard of the Kolmogorov-Smirnov test ($p > 0.05$) (Tab. 3.3.6). The research data meet the conditions of asymmetry and excess – the asymmetry and excess coefficients range from −1 to +1. Bearing in mind that the conditions of asymmetry and excess are met, it is possible to claim that the normality standard are fully complied with and ANOVA data block test can be applied.

Applying ANOVA data block test, the statistical significance of the difference between the means of IMI subscale data is analysed in accordance with the criterion of Sphericity Assumed. It confirms the statistical significance of the difference in the data ($p = 0.000$), when the level of significance is $\alpha = 0.05$ (Tab. 3.3.7).

The greatest statistically significant differences were determined among the evaluation means of the first scale *interest/enjoyment* and five subscales (*2. Feeling of competency; 3. Effort; 4. Felt tension and pressure; 6. Value/usefulness; 7. Relatedness*) ($p > 0.05$) (Tab. 3.3.7). Therefore, the feeling of intrinsic motivation of students is highly evident while performing Physics labs at the guided inquiry level. It statistically significantly exceeds the feeling of competency, effort put into the accomplishment of a lab, tension (pressure) in education and **relatedness**.

A statistically insignificant difference was detected between the first scale *interest/enjoyment* and the fifth scale *perceived choice (autonomy)*. Consequently,

Tab. 3.3.7: Results of ANOVA Bonferoni data block test at guided level Physics labs. Pairwise comparisons between the main factors of Self-Determination Theory: 2 – feeling of competency; 5 – perceived choice or autonomy; 7 – relatedness

(I) factor1	(J) factor1	Mean Difference (I–J)	Std. Error	Sig.[a]	95 % Confidence Interval for Difference[a]	
					Lower Bound	Upper Bound
Interest/ enjoyment	2. **Feeling of competency**	2.104	1.325	0.000	−2.022	6.230
	3. Effort	6.491*	1.368	0.000	2.232	10.749
	4. Felt pressure and tension	10.443*	1.515	0.000	5.727	15.159
	5. **Perceived choice (autonomy)**	4.066	1.813	0.567	−1.578	9.711
	6. Value/ usefulness	−11.858*	1.828	0.000	−17.551	−6.166
	7. **Relatedness**	9.236*	1.398	0.000	4.882	13.590

the students' intrinsic learning motivation while accomplishing Physics labs following the guided inquiry methodology bears no statistically significant difference from their feeling of ***perceived choice (autonomy)*** (Tab. 3.3.7).

Physics labs performed according to the guided inquiry methodology promote the students' feeling and understanding of the usefulness of experimental activity. The values of this subscale statistically significantly surpass the values of the subscale *interest/enjoyment* (Tab. 3.3.7).

A detailed analysis of the intrinsic motivation compliant with the subscale *interest/enjoyment* was carried out: *I liked doing this activity; It was fun doing it; I thought such an activity was boring; This activity did not attract my attention at all; I would define this activity as a very interesting one; I thought this activity was quite enjoyable; When doing this activity I was thinking how enjoyable it was.*

The research data demonstrated (Tab. 3.3.8) that 44.00 % of the respondents attributed the highest evaluation (5 – very likely) to the first question of the scale *(I liked doing this activity).* Therefore, it is possible to claim that the students considered the activity performed in guided inquiry Physics labs as interesting and enjoyable. 50.5 % of the students gave the highest evaluation to the second question of the scale (*It was fun doing it*). 37.4 % of the students evaluated the activity as an interesting one by choosing the fourth rank of evaluation (4 – more likely than unlikely); slightly fewer students – 36.3 % evaluated this activity by

Tab. 3.3.8: Frequencies of the variables of the students' intrinsic motivation for learning on the subscale *interest/enjoyment* (%): guided inquiry Physics labs

Rank of evaluation	Interest/ enjoyment	Feeling of competency	Effort	Felt pressure/ tension	Perceived choice (autonomy)	Value/ usefulness	Relatedness
4 – more likely than unlikely	25.3	19.8	3.3	3.3	37.4	30.8	15.4
5 – very likely	44.0	50.5	6.6	9.9	36.3	27.5	23.1

giving the highest rank (5 – very likely). Hence, the students considered this non-traditional activity as interesting due to the possibility to decide upon the procedure of the work and business-like communication in experimental groups. The sixth questions of the scale *I thought this activity was quite enjoyable* was also answered positively by the respondents: 30.8 % of them selected the fourth rank of evaluation (4 – more likely than unlikely). Hence, it is assumed that the feeling of interest/enjoyment of the activity while performing guided inquiry Physics labs is relatively high.

The research data (Tab. 3.3.8) revealed that the students found Physics labs that were performed in accordance with the guided inquiry methodology interesting as they did not know the inquiry procedure in advance, which demanded more creativity and promoted the students' interaction, search for information, and cooperation. The result of the investigation was not known, either, which in its turn also stimulated the students' motivation for learning Physics.

The correlation between the data of different subscales was also explored (Tab. 3.3.9). According to SDT, humans are actively seeking for the actualisation of their potentialities fulfilling their basic psychological needs: needs for autonomy, competency, and social relatedness. It was important to determine how the data of the first (*interest/enjoyment*) subscale correlated with the data of the second (*feeling of competency*), the fifth (*perceived choice*) and the seventh (*interpersonal interactions*) subscales at guided inquiry Physics labs. The strongest statistically significant correlation was determined between the students' intrinsic motivation for learning Physics and the feeling of competency (the first and the second subscales) ($r = 0.569^{**}$, $p = 0.01$). Hence, guided inquiry Physics labs gave the students' the feeling of competency and promoted their positive motivation for learning Physics.

The weakest and statistically insignificant correlation was detected between the students' intrinsic motivation (*interest/enjoyment*) and *perceived choice/ autonomy* ($r = 0.031$, $p = 0.05$). A weak yet statistically significant correlation was detected with the third psychological need – *relatedness* ($r = 0.269^{**}$, $p = 0.01$).

Tab. 3.3.9: Promotion of the students' intrinsic motivation for learning Physics through guided inquiry Physics labs: Pearson correlation coefficients between IMI questionnaire subscales

	Interest/ enjoyment	**Feeling of competency**	**Effort**	**Felt pressure and tension**	**Perceived choice (autonomy)**	**Value/ usefulness**	**Relatedness**
Interest/ enjoyment	1	,569**	,289**	,158	,031	,594**	,269**
Feeling of competency		1	,353**	,174	,131	,470**	,160
Effort			1	,232*	,319**	,458**	,282**
Felt pressure and tension				1	,273**	,040	,165
Perceived choice (autonomy)					1	,037	,135
Value/ usefulness						1	,093
Relatedness							1

** Correlation is significant at the 0.01 level (2-tailed).
* Correlation is significant at the 0.05 level (2-tailed).

3.3.3.3 Comparison of school students' intrinsic motivation at structured and guided inquiry Physics labs

The research results on the performance of Physics labs at the levels of structured and guided inquiry levels were compared according to the seven scales of IMI questionnaire: *interest/enjoyment' perceived competency' effort' value/usefulness' felt pressure and tension' perceived choice (or autonomy of activity)*, and *relatedness* (Tab. 3.3.10). Levene's Test demonstrated that in all compared cases $p > 0.05$. It means that the variances of the populations from which different samples were drawn were equal. Levene's test was used to assess this assumption. Thus, the null hypothesis of equal variances was confirmed and it was concluded that there was no difference between the variances in the population in terms of the different subscales of IMI.

According to SDT, the main IMI subscales are the following: *interest/enjoyment' perceived competency' perceived choice (or autonomy of activity)*, and *relatedness*. Statistically significant differences were detected in only one scale, namely *feeling of competency* (Tab. 3.3.10). The inquiry procedure which was not known in advance in the case of guided inquiry weakened the students' feeling of

Tab. 3.3.10: Comparison of the obtained data on structured and guided inquiry Physics labs performed in terms of IMI subscales: results of t-test

	Levene's Test for Equality of Variances		**t-test for Equality of Means**						
	F	**Sig.**	**t**	**df**	**Sig. (2-tailed)**	**Mean Difference**	**Std. Error Difference**	**95 % Confidence Interval of the Difference**	
								Lower	**Upper**
Interest/ enjoyment	0.068	0.795	1.512	205	0.132	2.673	1.768	–.813	6.160
Feeling of competency	0.379	0.539	2.439	205	0.016	5.728	2.349	1.097	10.359
Effort	7.653	0.006	0.307	205	0.759	0.520	1.695	–2.823	3.862
Felt pressure and tension	0.621	0.432	1.197	205	0.233	2.008	1.677	–1.298	5.314
Perceived choice (autonomy)	0.506	0.478	1.010	205	0.313	–2.142	2.120	–6.321	2.038
Value/ usefulness	0.478	0.490	–0.794	205	0.428	–2.571	3.237	–8.954	3.811
Relatedness	2.913	0.089	1.926	205	0.055	3.246	1.685	–.076	6.568

competency as they felt less competent (Tab. 3.3.5) than in the case of structured inquiry (Tab. 3.3.1). In the case of other subscales *relatedness* and *perceived choice (autonomy)*, a statistically insignificant difference was detected while performing Physics labs at both guided and structured inquiry levels.

In the case of the first subscale *interest/enjoyment*, the obtained data revealed a statistically insignificant difference between guided and structured Physics labs ($p = 0.132$), the significance level being $\alpha = 0.05$. Therefore, it is possible to claim that both guided and structured inquiry Physics labs encourage the students' intrinsic motivation for learning.

The employed Student t criterion demonstrated a statistically insignificant difference ($p = 0.759$) between guided and structured inquiry levels in terms of the scale *effort/importance*. Hence, both structured and guided inquiry Physics labs demanded equal proportion of effort to perform them. Analysing the psychological well-being of the students while performing Physics labs, a statistically insignificant difference was determined in terms of the scale *tension/pressure* ($p = 0.233$). Thus, in terms of the creation of a favourable psychological well-being of the students, guided and structured inquiry Physics lab are tantamount.

A statistically insignificant difference was detected in terms of the subscale *autonomy* ($p = 0.313$), and *relatedness* ($p = 0.055$) (Tab. 3.3.10). In both cases – structured and guided Physics labs, the students' autonomy of activity and the feeling of social interaction differed statistically significantly.

A statistically insignificant difference was also detected in the subscale *value/usefulness* ($p = 0.428$). The mean of the values of this subscale was the highest at the structured and guided inquiry levels (Tab. 3.3.1; Tab.3.3.5). Hence, the students' recognised the value of Physics labs and the practical applicability of the acquired abilities. This feeling was dependent on neither level of Physics labs: neither on structured nor guided inquiry.

3.3.4 Conclusions

The results of the research showed that the students did find Physics lab activity enjoyable, interesting, useful, and related to perceived choice, relatedness and feeling of competency. Physics labs are useful for the actualisation of students' potentialities fulfilling their basic psychological needs: needs for autonomy, competency, and social relatedness. Students understand the value and usefulness of Physics labs. They are intrinsically motivated and engage in Physics lab activity, since they understand the value and usefulness of Physics labs. In real Physics labs, the new generation students fulfilled the need for competency. It means that they fulfilled the desire to control and master the laboratory environment and outcomes of Physics labs. The results of the research revealed that the need for competency was best satisfied in real Physics labs, whereas the need for relatedness with other students to interact with and to be connected to other students of Physics labs was satisfied at a lesser degree. Physics labs satisfied the need of autonomy of students the least. Autonomous does not mean that students are independent at Physics lab activity. They act according to school programmes. Autonomy means having a sense of free will when doing something of own interests and values. The results of the research showed that students had at least satisfied this sense.

The perspectives of inquiry levels perspectives enabled us to understand what was happening during the structured inquiry and guided inquiry Physics labs at school. The three characteristics represent the areas in Physics labs where students could act independently: 'problem/question'; 'procedures/design'; and 'Conclusions'. At the level of structured inquiry, the students found out the problem and procedures of Physics labs. The conclusions were not known before Physics labs. At the level of guided inquiry, the students were provided with the problem and procedures of Physics labs but the methods of analysis, communication, and conclusions were for the student to design themselves. The means of the data of the subscales interest/enjoyment, feeling of competency, and effort were higher at the level of structured inquiry Physics labs. But only in one case a statistically significant difference was detected, namely on the subscale of feeling of competency. At the level of structured inquiry, the students acted less independently. It required less competency and the students felt more competent at less complex situations.

In Physics labs at the guided inquiry level, the means of the data of the subscales perceived choice (autonomy), and value/usefulness were higher as compared to the structured inquiry Physics labs. However, this difference was not confirmed statistically. In the guided inquiry labs, the students acted more independently while selecting the methods of analysis, communication, and formulating conclusions. It caused a sense of autonomy and usefulness of Physics labs.

3.4 Socio-scientific issues and communication in science education

By Palmira Pečiuliauskienė

The section deals with the role of socio-scientific issues (SSI) in promoting intrinsic motivation and communication of secondary school students. Their intrinsic motivation is analysed on the basis of SDT. The communication of school students was analysed according to three levels of communication: I-It; I-You; I-Thou (Buber, 1970). The research is based on disclosing the relationship between intrinsic motivation for learning sciences and different levels of communication: I - It; I - You; and I - Thou.

The relationship between the socio-cultural environment and motivation of a learner is reflected in the description of motivation: "motivation can be seen as a result of interactions between an individual's need system and the environmental factors that interfere with or support the need fulfilment process" (Deci & Ryan, 2008, p. 183). Communication is central to the interaction with a social context. Though educational practice is based on subject-specific communication, it would be impossible without interpersonal communication (Peciuliauskiene, 2015).

Research and innovations change the socio-cultural environment. Research has traditionally been seen as a moral and public good. The growing power of technology demonstrates the downsides of research and innovation. This situation reveals the intrinsic uncertainty of the future consequences of research and innovation (Stahl, 2013). A responsible research and innovation (RRI) is a new concept which describes the new situation of research and innovation. According to Von Schomberg RRI is "... a transparent, interactive process by which societal actors and innovators become mutually responsive to each other with a view on the (ethical) acceptability, sustainability and societal desirability of the innovation process" (Von Schomberg, 2011, p. 9). RRI needs to engage with the possible consequences of research and innovation through foresight and assessment processes (Stahl, 2013; Sherborne, et al., 2014). Students should have access to realistic information about science and technologies through direct contact with professionals (Loukomies et al., 2013). The partnerships between schools and industrial organisations are important for students' engagement (Tytler et al., 2008). It means that RRI "requires a dialogue between scientists and all citizens including the youth to better align the results of research with societal needs" (Okada, 2016, p. 11). How does the dialogue between scientists and students engage them in science?

Communication is central to collaborative work, as well as to the interaction with a social context. The teamwork approach offers a vehicle for the transfer of the control of learning from a teacher to learners, encourage students' responsibility for learning and promote intrinsic motivation (Lawlor et al., 2015). Individual team members enjoy the support of their peers and shared responsibility (Pyle, 1995).

Students' intrinsic motivation is influenced by the level of collaborative work, the relationship among peers, teacher-student relationship, and the social environment. RRI enables to focus on intrinsic motivation by extending education areas from formal to informal education, and by extending communication from inter-student relations or teacher-student relations to student-scientists. RRI requires a dialogue between scientists and other societal actors. Students must be equipped to work as qualified professionals responsible for innovation that is desirable, acceptable, and sustainable. Establishing communication between scientists and non-scientists during research is not easy. Trust and communication abilities are essential for a proper dialogue between societal actors. Inquiry activities with various societal actors in formal and informal scenarios open up new opportunities for science education at schools and beyond the schools (Okada, 2013).

SDT is one of the leading theories for understanding human motivation (Byman et al., 2012). The main idea of SDT is that a person is in interaction with social contexts (Reeve et al., 2004). Humans do not live in vacuum as their social environments can either facilitate or inhibit these inherent tendencies (Ryan & Deci, 2002; Loukomies et al., 2013). A key challenge for teachers is to change how science is taught by moving from content-based lessons to lessons based on the social environment content, as well as by moving to skills and values to use science in the society (Okada, 2016). Science classrooms should be promoting "scientific thinking in a community of scientists" because it engages students more deeply in science and moves away from learning of isolated facts (Nolen, 2003, p. 349). Students are engaged in science through various forms of active learning (Healey & Roberts, 2004; Healey, 2005; Igwebuike & Oriaifo, 2012).

SSI engages school students in critical discussions on current scientific topics and moves away from learning of isolated facts. SSI are socially controversial (or socially alive) topics or issues which have a scientific component but also incorporate other disciplines and interests (political, economic, ethical, etc.) and which involve the evaluation of moral and ethical aspects (Evagorou, Jimenez-Aleixandre, & Osborne, 2012). Von Schomberg (2011) states that SSI requires learners to be able to deal with uncertainties, as well as to better understand the potential benefits and risks of science discovery.

In the related literature, there has been an emphasis in the study of SSI regarding students' decision making and conceptual understanding (Zeidler & Nichols, 2009; Espeja & Lagarón, 2015). This situation necessitates for a deeper look into the problem of SSI at secondary school - by the aspect of intrinsic motivation for learning science of secondary school learners.

The discussed situation highlights the scientific problem which is formulated as a question: what is the interpersonal communication of students at the SSI activity and how is it related to intrinsic motivation for learning science?

The object of the research is the intrinsic motivation for learning science and interpersonal communication of students.

The aim of the research is to reveal the relationship between intrinsic motivation of students towards science and their interpersonal communication at SSI activity.

The objectives of the research are as follows:

1. How does SSI activity engage school students in science?
2. How is the intrinsic motivation towards learning science of students related to the levels of interpersonal communication at SSI activity?

3.4.1 Method of research

The data presented in the current research are a part of the ENGAGE project in Lithuania. Inquiry-based learning and RRI approach were used in the experimentation activity of the eighth-tenth grade students. Addressing the problem solving tool, the 5E instructional model was employed to show how to use it to teach an enquiry process and apply science content in a contemporary issue (Boddy et al., 2003; Bybee et al., 2006), using the materials created during the

ENGAGE project. The model consists of a sequence of five learning phases – Engage, Explore, Explain, Elaborate, and Evaluate. Each phase has a specific function and contributes to the intrinsic motivation and interpersonal communication. These phases correspond to different steps of the five steps method. In the 5E model, the curriculum is designed to allow students to explore scientific phenomena and their own ideas. Students are invited to explain their ideas, and explanations can also be provided by a teacher, a textbook, or through the usage of technologies. Students are able to explore, explain, extend, and evaluate their own progress. The teamwork approach was chosen. Students worked in groups of four in one group.

The instrument of quantitative research. IMI (Ryan, 1982; R'an, Koestner, & D'ci, 1991) was used when carrying research. IMI has seven subscales, while the first subscale (called interest/enjoyment) and the seventh subscale (called relatedness) were used during this research. Participants' communication during laboratory experiments was assessed using the seventh subscale of IMI. This subscale characterizes learners' thoughts and feelings regarding another person participating in the experiment. The interval scale (range from 1 to 100 points) was used to represent the results of the seventh subscale in our research.

The sample and sampling of quantitative research. We organised on-line courses of 6 weeks (19 October–30 November, 2015) for science teachers in order to appraise how contemporary science can engage students and get them think and talk, to master the use of a 5E lesson to develop RRI/inquiry skills, and to be able to design your own problem-based lessons for RRI/enquiry skills. 47 science teachers completed the on-line courses. They tried out the obtained theoretical materials in the classrooms with their students. The sample included sixth-tenth form students of different Lithuanian schools. The research clusters were the largest cities of Lithuania. The classes were selected on the basis of probability cluster sample, and all the learners of a selected class have been tested.

There are 25,000 eighth-tenth form school students in Lithuania (EMIS). The confidence interval of 5 % and confidence level of 95 % was chosen (the sample should be composed of at least 379 respondents). The sample of our research involves 400 students showing the reliability of the sample.

3.4.2 Results

Exploratory factor analysis (EFA) was used for investigating variables of motivation. EFA complies with the research aim because it reduces data to a smaller set of summary variables and to explore the underlining theoretical structure of the phenomena. The factorability of the 7 items (*1. I enjoyed doing this activity very*

Tab. 3.4.1: Rotated component matrix of observed intrinsic motivation variables: factor loading.

Observed variables	Latent variables: factors	
	F1 (Engagement)	**F2 (Humdrum)**
I would describe this activity as very interesting	0.847	
This activity was fun to do	0.832	
I thought this activity was quite enjoyable	0.809	
I enjoyed doing this activity very much	0.754	
While I was doing this activity, I was thinking about how much I enjoyed it	0.686	
This activity did not hold my attention at all		0.903
I thought this was a boring activity		0.833

much; 2. This activity was fun to do. 3. I thought this was a boring activity; 4. This activity did not hold my attention at all; 5. I would describe this activity as very interesting; 6. I thought this activity was quite enjoyable; 7. While I was doing this activity I was thinking about how much I enjoyed it) of intrinsic motivation (the first IMI subscale) was examined. The skewness and kurtosis were well within a tolerable range for assuming a normal distribution. The Kaiser-Meyer-Olkin measure of sampling adequacy was 0.792, above the commonly recommended value of 0.6, and Bartlett's test of sphericity was significant (χ^2 (21) = 1,236,820, $p < 0.05$). The diagonals of the anti-image correlation matrix were also all over 0.5.

Principal component analysis (PCA) was used for the computation of the scores for the factors underlying the subscale of intrinsic motivation. Initial eigenvalues indicated that the first three factors explained 50.865 %, 18.414 %, and 10.682 % of the variance, respectively. Solutions for two or three factors were each examined using varimax rotations of the factor loading matrix. Varimax rotations revealed that two factors explained 69 % of the variance with the factor extraction via eigenvalues more than 1.

Factor loadings were represented in the rotated component matrix (Tab. 3.4.1). The relationship of each variable to the underlying factor was expressed by factor loading. The names of the factors were Engagement (the first factor) and Humdrum (the second factor).

RRI enables different levels of communication. Communication varies from impersonal to interpersonal communication (Wood, 2013). Three levels of communication are set by Buber (1970): I-It; I-You; I-Thou. The lowest level was named as I-It representing that other people are treated very impersonally, without acknowledging the humanity of them and treating them only as objects.

Tab. 3.4.2: Spearman correlation coefficients between student's engagement in RRI activity and interpersonal communication: I–Thou and I–You level

Variables of the factor *Engagement*	**Variables of factors**			
	I–You		**I–Thou**	
	It is likely that this person and I could become friends if we interacted a lot	**I'd like a chance to interact with this person more often**	**I felt like I could really trust this person**	**I feel close to this person**
I would describe this activity as very interesting	0.259**	0.179**	0.190**	0.212**
This activity was fun to do	0.333**	0.300**	0.248**	0.171**
I thought this activity was quite enjoyable	0.320**	0. 268**	0. 227**	0.112*
I enjoyed doing this activity very much	0.241**	0.190**	0.118*	0.149*
While I was doing this activity, I was thinking about how much I enjoyed it	0.315**	0.231**	0.209**	0.175*

** Correlation is significant at the 0.01 level (2-tailed).
* Correlation is significant at the 0.05 level (2-tailed).

The higher level of communication was named as I-You. At this level people acknowledge others more than objects, others are recognised as individuals within those roles but still full engagement with each other is missing. The interaction is predetermined by social roles of people (members of a team, classmates, colleagues, etc.) (Buber, 1970). "In the workplace, most of us have many I-You relationships"(Wood, 2013, p. 19). The highest level of communication is I-Thou where people completely open themselves and totally accept others (Buber, 1970).

In order to disclose the relationships between interpersonal communication at RRI activity and intrinsic motivation of students, correlation analysis was completed. Spearman (r) was used for the correlation analysis because two variables being analysed were measured on a range scale. Spearman correlation coefficients (r) were calculated between the variables from the factor *Engagement* and variables from the factor I-Thou and I-You (Tab. 3.4.2). All correlations between the two variables were positive (i.e. higher levels of one variable were associated with higher levels of the other). The magnitude of the correlation coefficient indicated the strength of the association. In our case, the strength of association was weak (0.1–0.3) or moderate (0.3–0.5) (Tab. 3.4.2).

Four weak but statistically significant correlations between factors Engagement and I–Thou variables were revealed. One weak correlation between the variables *I feel close to this person* and *I would describe this activity as very interesting* (r = 0.212**) was disclosed. Three weak but statistically significant correlations between the variable *I felt like I could really trust this person* and the variables *This activity was fun to do* (r = 0.248**), *I thought this activity was quite enjoyable* (r = 0.227**), and *While I was doing this activity, I was thinking about how much I enjoyed it* (r = 0.209**) were revealed.

Five weak but statistically significant correlations between the factors Engagement and I-You were established (Tab. 3.4.2).

The correlations between the variables of the factor Engagement and variables of the factor I-You were stronger than between the variables of the factor I–Thou. Three statistically moderate correlations between the variables *It is likely that this person and I could become friends if we interacted a lot*, and the variables of the factor *engagement* (*This activity was fun to do* (r = 0.333**); *I thought this activity was quite enjoyable* (r = 0.320**); *While I was doing this activity, I was thinking about how much I enjoyed it* (r = 0.315**)) were calculated.

3.4.3 Conclusions

One of the important results of this study is that the school students were engaged in RRI activity implemented by Research-based way. Factor analysis revealed the group of variables related with intrinsic motivation for learning science on the basis of IMI scale. The learners pointed out that RRI activity is very interesting; quite enjoyable; this activity was fun to do. The factor loading of variables with factor Enjoyment are high: I would describe this activity as very interesting - 0.847; This activity was fun to do - 0.832; I thought this activity was quite enjoyable - 0.809.

The evidence from the data indicated statistically moderate relationship between participants' I-You variables and Engagement variables. The magnitude of the Spearman correlation coefficient indicated the strongest relationship between the variable *This activity was fun to do* (Engagement) and the variable *It is likely that this person and I could become friends if we interacted a lot* (I-You) (r = 0.333**) compared to other associations between the variables of the factor Engagement and variables of the factors I-You and I-Thou. The relationship between expectation of social relatedness (*I'd like a chance to interact with this* *person more often*) and engagement in science learning (*This activity was fun to do*) was expressed by Spearman correlation coefficient (r = 0.300**), which was only slightly lower than the correlation between engagement in learning science (*This activity was fun to do*) and expectation about friendship (*It is likely that this person and I could become friends if we interacted a lot*). The level of interpersonal communication I–You, which was based on expectation about friendship and social relatedness, made learning enjoyable and funny.

3.5 Student motivation for participation in Physics competitions

By Valdemaras Aleksa

The section analyses the role of non-formal Physics education in promoting motivation for learning Physics. The motivation of gifted students for learning Physics is analysed with reference to SDT, namely its sub-theory Organismic Integration Theory (OIT).

3.5.1 Introduction

It is a paradox that students' interest in nature sciences has been decreasing in most of European countries, including Lithuania, in the age of technologies (Dierking & Falk, 2016). Therefore, it is highly important that students gifted for sciences are motivated by organizing Olympiads, competitions, scientific events, and projects on different school subjects, as well as creating proper conditions for students to attend schools offering non-formal or part-time education.

In Lithuania, students gifted in Physics can additionally attend several schools offering part-time education, where they deepen the knowledge of Physics and Mathematics acquired at school. Besides, students interested in Physics can participate in a number of different events meant for the popularization of this field of science. One of the long-lasting events, which has been organized since 1990, is Physics Championship, where all willing school students of grades 9–12 can take part. The Championship is organized on the first Saturday in December every year in the largest cities of Lithuania: Vilnius, Kaunas, Klaipėda, Šiauliai, and Panevėžys.

Among other Physics competitions, the Physics Championship is distinguished by the fact that all participants receive the same set of assignments that is composed of 10 assignments related to mechanics, molecular physics, electricity, and optics. The Physics Championship is organised by Lithuanian Centre of Non-formal Youth Education, whereas the assignments are prepared by teachers of different Lithuanian universities with reference to the programme of Physics Olympiads that is in tune with the general school curriculum for Physics. Each assignment is assessed by 10 points; hence a student can collect 100 points after successful solution of all the assignments. Considering the number of collected points, prize places are awarded to classes, whose three best students have collected the biggest amount of points. The participant that has collected the greatest number of points of all is awarded the title of Physics Champion and is presented with the Champion's Throne, a special chair that is traditionally established by the sponsor. The winners of the 1st–3rd place are awarded diplomas signed by the Minister of Education and Science. The winners of the 1st and the 2nd places are invited to participate in the Physics Olympiad in Lithuania, whereas the best five students that succeed at the national level are offered to participate in the international Physics Olympiad, in which Lithuanian students participate since 1989.

Annual Physics championships have become a tradition in Lithuania, yet the number of students has been constantly decreasing: in comparison to 2004, when 1246 students participated in the championship (the record number), in 2016 the number was three times smaller (according to the data of Lithuanian Centre of Non-formal Youth Education). The analysis of the number of the participants reveals that it decreased to 390 students from 2006 to 2016. Why does this happen? On the one hand, the reduction in the number of participants can be affected by the demographic situation of the country; on the other hand, it should be noted that the average results demonstrated by the participants of each age group (grades 9–12) have not decreased and a positive tendency is even observed.

The current research aims at answering the following question: what motifs stimulate students' self-determination to learn Physics and participate in the Physics Championships? Motivation is one of the most significant factors. Learning is affected by both extrinsic and intrinsic motivation. Extrinsic motivation is an individual's need to accomplish a particular activity that is determined by the environment. It involves a person's understanding of what he/she wants to achieve and what they should do for a particular reason; yet the motifs for such activities are not determined by students themselves – they can be conditioned by parents' expectations, influence of respectful or important people, an opportunity to get a reward or other benefits. Therefore, external motivation is short-term – if the stimulating factor is eliminated, a person loses motivation to perform the activity. Intrinsic motivation is an individual's internal need for performing an action. Person having intrinsic motivation clearly understand their interests, enjoyable activities and seek to perform them themselves. Therefore, intrinsic motivation is stronger than extrinsic; it is long-lasting and self-contained, which helps to achieve better results. Intrinsic motivation is stimulated by the freedom of choice, tasks correspondent to students' abilities, an opportunity to contribute to solving a genuine problem, examples of other people, as well as the feeling of belonging (Reeve, 2002; Reeve & Jang, 2006). Intrinsic motivation increases when external and internal motifs are related to a person's interests.

OIT (Deci & Ryan, 1985; Deci & Ryan, 2000) has been employed to explore students' external and internal motifs to participate in the Physics Championship and to reveal the consistency of their motivation. The aforesaid theory focuses on the diversity of overlaps between learning, social and self-contained motifs in the continuum from extrinsic to intrinsic motivation.

The 28th Lithuanian Physics Championship organised for school students in 2016 was employed in the research.

The object of the research is students' motivation to participate in the Lithuanian Physics Championship.

The aim of the research is to explore the main motifs of students determining their participation in the Lithuanian Physics Championship organised for school students.

The research objectives are as follows:

1. To disclose the motifs of 9th–12th grade students to participate in the Physics Championship highlighting the differences in motivation.
2. To identify the compliance of students' motifs to participate in the Physics Championships with the levels of OIT motivation in the continuum from extrinsic to intrinsic motivation.

The hypothesis of the research. The Physics Championship, a non-formal educational event, stimulates the self-determination of students gifted in sciences to learn Physics and link their professional career to Physics.

3.5.2 Method of research

Students' motivation for participating in the Physics Championship was explored using questionnaire survey that was performed online by emailing a link to the questionnaire for the participants of the 28th Championship of 3 December 2016. 390 participants solved the Physics Championship assignments; 9 out of them did not indicate their personal e-mails. Hence, the questionnaires were sent out to 381 Championship participants on 1 March 2017, i.e. 19 days after the announcement of the results of the 28th Physics Championship. 132 students (34.6 %) filled in the questionnaire.

The questionnaire was developed considering the specifics of the Physics Championship. The students were asked the following questions: 1) why did

Tab. 3.5.1: Characteristics of students attending the Physics Championship

Grade	Age/Date of Birth	Number of respondents exact number/percentage
9th	15/2002	43/32.6 %
10th	16/2001	34/25.8 %
11th	17/2000	31/23.5 %
12th	18/1999	24/18.2 %

you decide to take part in the Physics Championship? 2) What were your motifs? 3) How did you benefit from the participation? The current work analyses the students' responses to the second question – What were your motifs for participating in the Physics Championship? 5 answer rankings were provided (1-*Totally agree*, 2-*Agree*, 3-*Neither agree nor disagree*, 4-*Disagree*, 5-*Totally disagree*).

The obtained data on students' motifs for participating in the Physics Championship were evaluated qualitatively aiming to reveal the logic between external and internal motifs and their compliance with the levels of motivation identified by OIT sub-theory: *External Regulation*, *Introjected Regulation*, *Identification*, *Integrated Regulation*, *Intrinsic Regulation* (Deci & Ryan, 1985; Deci & Ryan, 2000).

3.5.3 Results

Out of 132 respondents, the greatest number that filled in the questionnaire were the students of grade 9 (32.6 %), the least number were students of grade 12 (18.2 %) (Tab. 3.5.1). These numbers comply with the general tendencies of the distribution of students in Physics Championships. The majority of the respondents claimed having participated in the Physics Championship for the first time (56.8 %) and only 7 students noted that they participated for the fourth time – 5.3 % of the total number of the respondents.

The main research data were obtained from the matrix of the respondents' replies to the question "What were your motifs for participating in the Physics Championship?" and the five answer rankings (1 – *Totally agree*, 2 – *Agree*, 3 – *Neither agree nor disagree*, 4 – *Disagree*, 5 – *Totally disagree*). The data analysis concerning the respondents from grades 9, 10, 11, and 12 are presented in Tables (3.5.2–3.5.5), respectively.

The majority of **43 students of grade 9** maintained that the greatest motif for their participation in the Physics Championship was the science of Physics: *I like*

Tab. 3.5.2: The motifs of the 9th grade students to attend the Physics Championship

Motifs	**Answer rankings**				
	Totally agree	**Agree**	**Neither agree nor disagree**	**Disagree**	**Totally disagree**
I like Physics and I want to attend the Physics Championship	27	10	4	0	1
I want to learn Physics better/ learn something new	24	19	0	0	0
I want to get acquainted with Peers interested in Physics	4	15	17	7	0
I want to compete with others	11	17	5	9	1
I want to check/evaluate my knowledge	21	19	2	1	0
I want to win	8	14	15	6	0
I want to get a good mark in Physics /Increase the mark of Physics at school	6	8	14	4	11
My Physics teacher wants it	7	12	19	4	1
My parents want me to participate in the Physics Championship	2	4	14	16	7

Physics and I want to attend the Physics Championship (*Totally agree* – 27, *agree* – 10) and *I want to learn Physics better/learn something new* (*Totally agree* – 24, *agree* – 19) (Tab. 3.5.2). Slightly fewer students selected a different variant – motif *I want to check/evaluate my knowledge* (*Totally agree* – 21, *Agree* – 19), which was also related to motivation for learning Physics. An insignificant part of the respondents did not agree with the statements provided: *I like Physics and I want to attend the Physics Championship* (*Totally disagree* – 1) and *I want to check/evaluate my knowledge* (*Disagree* – 1), whereas 4 did not have a clear opinion (*Neither agree nor disagree*).

The students of this age group *wanted to compete with others* (*Totally agree* – 11, *agree* – 17) and *wanted to win* (*Totally agree* – 8, *agree* – 14); having summed up the options *Totally agree* and *agree* (44.2–72.1 %), it is possible to claim that every second student of grade 9 considers winning as important. The remaining students disagreed that they were motivated to participate in the Physics Championship by *willingness to compete* (9) and *willingness to win* (6).

Tab. 3.5.3: The motifs of the 10th grade students to attend the Physics Championship

Motifs	**Answer rankings**				
	Totally agree	**Agree**	**Neither agree nor disagree**	**Disagree**	**Totally disagree**
I like Physics and I want to attend the Physics Championship	17	13	3	1	0
I want to learn Physics better/ learn something new	19	13	0	1	1
I want to get acquainted with Peers interested in Physics	6	8	9	6	3
I want to compete with others	11	8	10	2	3
I want to check/evaluate my knowledge	19	14	1	0	0
I want to win	11	6	13	3	1
I want to get a good mark in Physics /Increase the mark of Physics at school	2	6	12	7	7
My Physics teacher wants it	7	10	8	4	5
My parents want me to participate in the Physics Championship	1	3	12	8	10

Every third student of grade 9 (15/34.9 %) agreed (*Agree*) with the motif-statement *I want to get acquainted with Peers interested in Physics* that was considered as important by the youngest respondents.

32.6 % of the respondents agreed with the statement *I want to get a good mark in Physics/Increase the mark of Physics at school* (*Totally agree* 6, *agree* 8). However, every fourth student *totally disagreed* (7) with the aforesaid statement.

A great part of the students of grade 9 maintained that they participated in the Physics Championship because of their teacher (*My Physics teacher wants it; Totally agree* – 7, *agree* – 12) and only a small part noted their parents' willingness (*My parents want me to participate; Totally agree* – 2, *agree* – 4). It is noteworthy that 53.5 % of the students disagreed with the above-mentioned statements (*My parents want me to participate: Disagree* – 16 and *Totally disagree* – 7), i.e. every second student of grade 9 believed that their parents had no influence on their decision to participate in the Physics Championship.

34 students of grade 10 filled in the questionnaire. The distribution of their replies regarding the motifs for their participation in the Physics Championship was similar to that of 9th grade students (Tab. 3.5.3). The greater majority of the respondents selected the answers related to learning Physics: *I want to check/evaluate my knowledge* (*Totally agree* – 19, *Agree* – 14), *I want to learn Physics better/learn something new* (*Totally agree* – 19, *Agree* – 13), *I like Physics and I want to attend the Physics Championship* (*Totally agree* – 17, *Agree* – 13). Nevertheless, 4 respondents claimed neither agreeing nor disagreeing to such statements (*Neither agree nor disagree)*, whereas 3 students claimed disagreeing with the aforesaid motifs: *I want to learn Physics better/learn something new* (*Totally disagree* – 1, *Disagree* – 1) and *I like Physics and I want to attend the Physics Championship* (*Disagree* – 1).

The motifs of competition and winning were also important in this age group. The motif *I want to compete with others* was considered as important by more than a half of the 10th grade students (*Totally agree* – 11, *Agree* – 8). Slightly fewer respondents marked the motif *I want to win* as important (*Totally agree* – 11, *Agree* – 6). However, 23 students of grade 10 expressed no clear opinion regarding this option (*Neither agree nor disagree); 3 totally disagreed* and 2 *disagreed* with the option *I want to compete*, whereas 1 student *totally disagreed*, and 3 students *disagreed* with the statement *I want to win.*

It is noteworthy that a half of the respondents of this age group maintained that they participated in the Physics Championship because their teacher wanted them to (*My Physics teacher wants it: Totally agree* – 7, *Agree* – 10). The remaining half of the respondents either *totally disagreed* – 5, *disagreed – 4, or neither agreed or disagreed – 8* with the statement.

Less than a half of the 10th grade students linked their motivation to participate in Physics Championships with their *willingness to get acquainted with peers interested in Physics* (*Totally agree* – 6, *Agree* – 8). Others found this motif as not relevant *(Totally disagree* – 3, *Disagree* – 6) or not important enough (*Neither agree nor disagree* – 9).

Almost a fourth of the students (8 students of grade 10) marked the statement *I want to get a good mark in Physics /Increase the mark of Physics at school.* Others expressed no clear opinion (*neither agree nor disagree* – 12, *disagree* – 7 and *totally disagree* – 7). A few of the 10th grade students (4) selected the motif *My parents want me to participate in Physics Championship* (*Totally agree* – 1, *Agree* – 3). 12 students *neither agreed nor disagreed*, whereas 8 *disagreed* and 10 *totally disagreed* with the statement.

31 students of grade 11 participated in the survey (Tab. 3.5.4). Within this group of the respondents, even a greater number of students indicated the motifs demonstrating their interest in the science of Physics: the statements *I like Physics and I want to attend the Physics Championship* (*Totally agree* – 18, *Agree* – 11), *I*

Tab. 3.5.4: The motifs of the 11th grade students to attend the Physics Championship

Motifs	Answer rankings				
	Totally agree	Agree	Neither agree nor disagree	Disagree	Totally disagree
I like Physics and I want to attend the Physics Championship	18	11	2	0	0
I want to learn Physics better/learn something new	16	12	2	0	1
I want to get acquainted with Peers interested in Physics	8	5	11	6	1
I want to compete with others	9	12	5	2	3
I want to check/evaluate my knowledge	19	10	2	0	0
I want to win	8	8	11	3	1
I want to get a good mark in Physics /Increase the mark of Physics at school	1	8	8	5	9
My Physics teacher wants it	8	12	6	4	1
My parents want me to participate in the Physics Championship	3	1	12	5	10

want to learn Physics better/learn something new (*Totally agree* – 16, *Agree* – 12), and *I want to check/evaluate my knowledge* (*Totally agree* – 19, *Agree* – 10) constituted a third of the total number of responses. 2 students had no clear opinion regarding each of the said statements (*Neither agree nor disagree*), whereas 1 replied that he/she *totally disagreed* with the statement *I want to learn Physics better/learn something new.*

More than a half of the 11th grade students claimed that they were motivated by their *willingness to compete with others* (*Totally agree* – 9, *Agree* – 12) and *to win* (*Totally agree* – 8, *Agree* – 8). However, 16 respondents *neither agreed nor disagreed* with the statement. 2 students *disagreed* and 3 *totally disagreed* with being motivated by the *willingness to compete with others;* whereas 3 *disagreed* and 1 *totally disagreed* with the statement *I want to win.*

8 respondents noted that they *totally disagreed* and 5 *agreed* being motivated by their wish *to get acquainted with peers interested in Physics* (41.9 %). 11 students *neither agreed nor disagreed with the statement*, 6 *disagreed* and 1 *totally disagreed* with the statement.

The motif *My Physics teacher wants it* was chosen by 64.5 % of the students from this age group (*Totally agree* – 8, *Agree* – 12). 1 respondent *totally disagreed*, 4 *agreed* and 6 *neither agreed nor disagreed* with the statement.

29.0 % of the total number of 11th grade students agreed with the statement *I want to get a good mark in Physics/increase the mark of Physics at school* (*Totally agree* – 1, *Agree* – 8). 8 students *neither agreed nor disagreed*, 5 *disagreed* and 9 *totally disagreed* with it.

Only a few respondents (12.9 %) replied that *their parents wanted them to participate in the Physics Championship* (*Totally agree* – 3, *Agree* – 1). 12 students noted that they *neither agreed nor disagreed*, 5 claimed that they *disagreed* and 9 maintained that they *totally disagreed* with the statement.

24 students of grade 12 filled in the questionnaire (Tab. 3.5.5). The respondents of this age group, like other age research participants, were mostly motivated to participate in Physics Championships by the subject of Physics: *I like Physics and I want to attend the Physics Championship* (*Totally agree* – 14, *agree* – 7) and *I want to learn Physics better/learn something new* (*Totally agree* – 11, *agree* – 9). Slightly more students selected the motif *I want to check/evaluate my knowledge* (*Totally agree* 13, *Agree* 11). 5 respondents *neither agreed nor disagreed* with the aforesaid statements, whereas 2 students *disagreed* with the statement *I want to learn Physics better/learn something new.*

A great part of the respondents (62.5 %) *wanted to compete with others* (*Totally agree* – 6, *agree* – 9) and half of them *wanted to win* (*Totally agree* – 3, *agree* – 9). 9 students expressed doubts (*neither agreed nor disagreed)* regarding these statements. 2 students *totally disagreed* with the respective statements, whereas 3 students *did not want to compete*, and 5 *did not seek to win*.

Fewer than half of the 12th grade students agreed with the statement *I want to get acquainted with peers interested in Physics* (*Totally agree* – 3, *agree* – 6). Almost the same number of the respondents (8) *neither agreed nor disagreed*, 6 *disagreed* and 1 *totally disagreed* with it.

Only 3 students of grade 12 claimed that they *I wanted to get a good mark in Physics /increase the mark of Physics at school*, a half of the respondents *totally disagreed* with the statement. 5 *disagreed* and 4 *neither agreed nor disagreed* with it.

One student *totally agreed* and 11 students *agreed* (50 %) that they *participated in the Physics Championship* because their *Physics teacher wanted;* whereas

Tab. 3.5.5: The motifs of the 12th grade students to attend the Physics Championship

Motifs	**Answer rankings**				
	Totally agree	**Agree**	**Neither agree nor disagree**	**Disagree**	**Totally disagree**
I like Physics and I want to attend the Physics Championship	14	7	3	0	0
I want to learn Physics better/learn something new	11	9	2	2	0
I want to get acquainted with Peers interested in Physics	3	6	8	6	1
I want to compete with others	6	9	4	3	2
I want to check/evaluate my knowledge	13	11	0	0	0
I want to win	3	9	5	5	2
I want to get a good mark in Physics /Increase the mark of Physics at school	0	3	4	5	12
My Physics teacher wants it	1	11	8	2	2
My parents want me to participate in the Physics Championship	0	0	9	6	9

no students claim that they participated in the Championship because of *their parents' wish.* The majority of the respondents expressed doubts regarding their participation in the Championship under the influence of their teachers and parents (17).

More students denied the statement *My parents want me to participate in the Physics Championship* (*Disagree* – 6 and *Totally disagree – 9)* than the statement *My Physics teacher wants it* (*Disagree* – 2 and *Totally disagree* – 2).

Analysing the motifs of new generation students gifted for Physics for attending the Physics Championship, the **most important replies that revealed motivation were considered 1 – *Totally agreed* and 2 – *Agree.*** The latter were chosen to identify the students' motivation.

Tab. 3.5.6: Distribution of 9th–12th grade students' motivation (in per cent) in accordance with the motifs listed in the questionnaire

Motifs	9th	10th	11th	12th
I like Physics and I want to attend the Physics championship	86.0	88.2	93.6	87.5
I want to learn Physics better/Learn something new	**100.0**	**94.1**	**90.3**	**83.3**
I want to get acquainted with peers interested in Physics	44.2	41.2	41.9	37.5
I want to compete with others	65.1	55.9	67.7	62.5
I want to check/evaluate my knowledge	93.0	97.0	93.6	100.0
I want to win	**51.2**	**50.0**	**51.6**	**50.0**
I want to get a good mark in Physics/Increase the mark of Physics at school	**32.6**	**23.5**	**29.0**	**12.5**
My Physics teacher wants it	44.2	50.0	64.5	50.0
My parents want me to participate in the Physics Championship	**13.9**	**11.8**	**12.9**	**0.0**

Aiming to evaluate the different numbers of the students of grades 9, 10, 11, and 12 (43, 34, 31, and 24, respectively) that filled in the questionnaires, the motifs of different grade students to participate in the Physics event were compared in percentage (Tab. 3.5.6). An obvious **difference in the motivation of 9th–12th grade students** was disclosed establishing correlations between the motif to *get a better mark in Physics/increase the mark of Physics at school* and the grade of students. It was determined that the higher the grade of students, the less they agreed with the statement that they were motivated by a possibility to get a better mark at school. A similar, yet less significant tendency was observed in the case of the statement *I want to learn Physics better/Learn something new:* all the students of grade 9 (100 %) agreed with the statement; however, in the case of 12th grade students, 83.3 % of the students marked this statement as important. Besides, the difference in motivation influenced by *parents* was highlighted: 12.9 % of the 11th grade students considered this motif as an important one; yet not a single 12th grade respondent agreed with this statement. No significant differences were determined among the choices of other motifs to participate in the Physics Championship.

Having summed up the replies of the respondents of all grades presented in Tables (3.5.2–3.5.5), the following priorities of motivation (*Totally agree* and *Agree)* were identified:

I want to check/evaluate my knowledge (126 responses; 95.5 %),

I want to learn something new (123; 93.2 %),
I like Physics and want to attend the Physics Championship (123; 93.2 %).
I want to compete with others (83; 62.9 %),
My Physics teacher wants it (68; 51.5 %),
I want to win (67; 50.8 %);
I want to get acquainted with peers interested in Physics (55; 41.7 %),
I want to get a good mark in Physics (34; 25.7 %),
My parents want me to participate in the Physics Championship (14; 10.6 %).

Aiming to broaden the overview of survey results, it is important to briefly present the students' replies to an additional open-ended question: *how did you benefit from the participation in the Physics Championship?* A few students mentioned material benefits of attending the Physics Championship, namely the possibility to *receive a diploma* or *a discount in non-formal education institutions.* Others maintained that they considered this event as important as it helped to prepare for *Physics examination, studies of Physics, and Lithuanian and international Physics Olympiads.* The latter motifs revealed that the Championship participants related their future profession to Physics.

The data of the Physics Championship (2016) organised for the 9th–12th grade students were analysed according to the *Self-Determination sub-theory* OIT model of motivation continuum from lowest levels of extrinsic motivation towards intrinsic motivation (see Fig. 1.1.2).

According to motivation levels defined in OIT (Rayan & Deci, 2000), the frequency of the choice of the respondents' replies was as follows (Tab. 3.5.7):

The motifs of Integrated Regulation (261 responses): *I want to learn Physics better/ learn something new, I want to compete with others, I want to get acquainted with peers interested in Physics*;
Identification (193): *I want to check/evaluate my knowledge, I want to win*;
Intrinsic Regulation (123): *I like Physics and I want to attend the Physics Championship*;
Introjected Regulations (82): *My Physics teacher wants it, My parents want me to participate in the Physics Championship*;
External Regulations (34): *I want to get a good mark in Physics /Increase the mark of Physics at school.*

3.5.4 Discussion

More than 80 countries participate in the annual International Physics Olympiad. The research on students' motivation to participate in the Lithuanian Physics

Tab. 3.5.7: The results of the analysis of the replies of all respondents (132) to the question *How did you benefit from the participation in the Physics Championship*?

Motivation levels	**Criteria**	**Motifs**	**Responses *Totally agree + Agree***		
			Distribution according to grades: exact number (per cent)	**Total. exact number (per cent)**	**Distribution according to the level of regulation. exact number**
External Regulation	Pragmatism is predominant. there are no values. Students are influenced merely by external motivation: encouragement. tangible rewards or punishments.	*I want to get a good mark in Physics*	Grade 9: 14 (32.2) Grade 10: 8 (23.5) Grade 11: 9 (29.0) Grade 12: 3 (12.5)	34 (25.8)	34
Introjected Regulation	Internal motivation appears. yet it is determined by external values. The opinion of other people is perceived as encouragement.	*My Physics teacher wants it*	Grade 9: 19 (44.2) Grade 10: 17 (50.0) Grade 11: 20 (64.5) Grade 12: 12 (50.0)	68 (51.5)	82
		My parents want me to participate in the Physics Championship	Grade 9: 6 (13.9) Grade 10: 4 (11.8) Grade 11: 4 (12.9) Grade 12: 0 (0)	14 (10.6)	
Identification	Students realise that Physics is beneficial to the students themselves. They adjust to external values. The opinion of another person loses its importance. personal benefit is perceived.	*I want to check/ evaluate my knowledge*	Grade 9: 40 (93.0) Grade 10: 33 (75.1) Grade 11: 29 (93.6) Grade 12: 24 (100.0)	126 (95.4)	193
		I want to win	Grade 9: 22 (51.2) Grade 10: 17 (50.0) Grade 11: 16 (51.6) Grade 12: 12 (50.0)	67 (50.7)	

Integrated Regulation	Learning of Physics enters the system of students' values and becomes part of their personality. Students do not think about personal benefits. yet the emotional component is not developed yet.	*I want to learn something new*	Grade 9: 43 (100.0) Grade 10: 32 (94.1) Grade 11: 28 (90.3) Grade 12: 20 (83.3)	123 (93.2)	261
		I want to compete with others	Grade 9: 28 (65.1) Grade 10: 19 (55.9) Grade 11: 21 (67.7) Grade 12: 15 (62.5)	83 (62..9)	
		I want to get acquainted with peers interested in Physics	Grade 9: 19 (44.2) Grade 10: 14 (41.2) Grade 11: 13 (41.9) Grade 12: 9 (37.5)	55 (41.7)	
Intrinsic Regulation	Students do not think about personal benefits; the process of identification begins. Internal values are predominant. Physics evokes positive emotions and intrinsic satisfaction.	*I like Physics and I want to attend the Physics Championship*	Grade 9: 43 (86.0) Grade 10: 30 (88.2) Grade 11: 29 (93.6) Grade 12: 21 (87.5)	123 (93.2)	123

Championship revealed that the students of grades 9–12 related their main motifs of participation primarily with the willingness to check their knowledge (*I want to check/evaluate knowledge*), as well as a wish to learn Physics better or learn something new. A considerably smaller group of motifs involved a wish to compete and win, to get acquainted with peers, as well as the advice received from a teacher. Among the least important motifs the students distinguished their willingness to get a good mark and parent's influence in their participation in the Championship. Comparing the correlation between the motif *to get a better mark in Physics/to increase the mark of Physics* at school and the grade of students, it was noted that only the students of younger grades considered this motif important; the parents' influence equalled to zero in grade 12. The obtained results confirmed the results of the research on the influence of Science Fairs and Olympiads carried out in the USA. According to the research, one of the most significant motifs was the aim to learn something new (Abernathy & Vineyard, 2001).

Discussing the results of the research, single answers of the students to the open-ended question *How did you benefit from the participation in the Physics Championship?* should be noted. Four students indicated that participation in the Physics Championship helped them *to prepare for Physics examination, studies of Physics, as well as Lithuanian and international Physics Olympiads.* These motifs are relevant in developing skills and abilities and at sustaining this motivation seeking to influence future career choices in the field of Physics. The aim of raising scientific motivation of the participants in Physics Olympiads has been recently discussed in Germany. The research reveals a tendency of the growing number of participants in Physics Olympiads in the country (Petersen & Wulff, 2017). Such school activities as Physics Championships are perceived as enrichment measures that are very promising in terms of promoting talented students.

There still is a demand for deeper knowledge of the effects of competitions on participating students seeking to develop strategies for the improvement of the education of gifted students. In the current research, the data on the motivation of the students participating in Physics Championship were analysed in accordance with the OIT model of motivation continuum from lower *Extrinsic Motivation* levels (*External® Introjected Regulation® Identification® Integrated Regulation*) towards *Intrinsic Motivation*.

The research results disclosed a relatively mature motivation of the students participating in Lithuanian Physics Championship for learning Physics. The research participants indicated the motif *I like Physics and I want to attend the Physics Championship* as the most significant. Out of 132 research participants, 123 students noted this motif, which means that they found it important to take part in an activity for the enjoyment and satisfaction inherent in engaging in the behaviour itself. According to OIT, it complies with the level of *Intrinsic Regulation.* The greatest frequency (261 responses) was determined at the level of *Integrated regulation* of external autonomous motivation continuum (*I want to learn Physics better/learn something new, I want to compete with others, I want to get acquainted with peers interested in Physics)*, when students' motivation to engage in the behaviour is congruent with one's sense of self, and when learning of Physics enters the system of their values and becomes part of their personality. Hence, it becomes the most self-determined form of external motivation.

The motifs *I want to check/evaluate my knowledge, I want to win* was noted by 193 respondents. They comply with the level of *Identification* and reveal a conscious acceptance of participation at the Championship as being important to achieve personally valued outcomes. The latter motifs are partly autonomous, since behaviour regulation is perceived as one's own regulation.

The students' motifs *My Physics teacher wants it* and *My parents want me to participate in the Physics Championship* are ascribed to the level of *Introjected Regulation*, as they are related to the behaviour stimulated by authorities – the students participate in the Championship as it is needed, beneficial and it raises their self-esteem. 82 students of different grades noted these motifs; however, the influence of these motifs on the students' motivation to participate in the Physics Championship was different: 51.5 % of the respondents were encouraged by teachers, and only 10.6 % were motivated by parents. As it could be expected, the motivation of 12th grade students to participate in the event was not influenced by parents.

34 students pointed out the motif *I want to get a good mark in Physics/Increase the mark of Physics at school.* It marks *external* regulation, in which a particular behaviour is well-reasoned and purposeful; yet it is not supported by internal values. Hence, the part of the students who participated in the Championship for pragmatic and external motifs was the least.

3.5.5 Conclusions

The research revealed three main motifs of the students of grades 9–12 of Lithuanian schools to participate in the Physics Championship. The largest group of motifs involves the motifs related to the students' willingness to check and improve their knowledge of Physics. The second group of motifs included willingness to compete and win, as well as to get acquainted with peers and satisfy the expectations of Physics teachers. The third least important group of motifs constitutes the motifs related to additional benefits at school, such as raising the mark and realisation of parents' expectation. Comparing the motifs of younger and senior school students, no significant differences were determined; however, a tendency of decreasing motivation to participate in the Physics Championship wishing to get more or new knowledge of Physics was estimated among senior students. Parents' expectations had no influence on the motivation of the 12th grade students.

The current research determined the compliance of the students' motifs to participate in the Physics Championship with the levels of motivation continuum from lower extrinsic regulation to intrinsic motivation as established by OIT: *External Regulations® Introjected Regulation® Identification® Integrated Regulation® Intrinsic Regulation.* The analysis of the frequency of responses, three main motifs of the students' participation in the Physics Championship that complied with the three regulation levels were identified: *Identification – I want to check/evaluate my knowledge; Integrated Regulation – I want to learn something new; Intrinsic Regulation – I like Physics and I want to attend the Physics Championship.* Having evaluated the frequency of motifs in the continuum, it was determined that most of the motifs were congruent with the level of *Integrated Regulation – I want to learn something new, I want to compete with others* and *I want to get acquainted with peers interested in Physics.* It is possible to claim that the motivation of the majority of the students gifted in Physics to participate in the Physics Championship is congruent with the sense of self, when the learning of Physics becomes a part of their identity and enters the system of values. Provided it is properly stimulated, such motivation can develop into the level of *Intrinsic Regulation*, when presumably Physics becomes an area of professional interest – *I like Physics and I want to attend the Physics Championship.* The research confirms the hypothesis that the Physics Championship, a non-formal educational event, stimulates the self-determination of students gifted in science to learn Physics and link their professional career to Physics.

4 Empirical insights into the motivation of school students for learning Mathematics

4.1 Extrinsic and intrinsic motivation of students for learning Mathematics

By Palmira Pečiuliauskienė

The section considers the motivation for learning Mathematics. The data of a national research (2012) on the 8th grade students are analysed according to the model of motivation continuum from the lowest level of extrinsic motivation to intrinsic motivation. Research is based on disclosing students' attitude towards the educational and social benefit of learning Mathematics.

4.1.1 Introduction

The theoretical background of motivation for learning is presented in the theoretical part of this monograph (The Theoretical Background of New Generation Student's Motivation for Learning). In this chapter we remind the main ideas of the Self-Determination (SDT) Theory on motivation in relation to the motivation for learning Maths. From the perspective of SDT, the three basic psychological needs are the need for competency (the desire to feel efficacious, to have an effect on one's environment, and to attain valued outcomes), the need for autonomy (the desire to be self-initiating and to have a sense of acting in accordance with one's own sense of self), and the need for relatedness (the desire to feel connected with and to be accepted by significant others) (Ryan & Deci, 2002; Loucomies, 2013).

Humans live in a socium; their social environments can either facilitate or inhibit motivation. Motivation can be seen as a result of interactions between an individual's need system and the environmental factors that interfere with or support the process of need fulfilment (Loucomies, 2013). Regulation of behaviour may be autonomous (self-determined) or controlled (Ryan & Deci, 2002). The sub-theories of SDT, for instance Organismic Integration Theory (OIT), describe motivation-like continuum and propose a taxonomy of regulation types ranging from non-regulation to intrinsic regulation. Along the continuum, internalisation of the regulation of behaviours increases and the motivation for certain behaviours becomes more autonomous (Ryan & Deci, 2002). The first point of motivation continuum is referred to as amotivation, which is based on non-regulation. The last point of the motivation continuum is entitled intrinsic motivation, and it is a prototype of autonomous and self-determined regulation. Extrinsic motivation based on controlled regulation occurs between amotivation and intrinsic motivation in the continuum (Fig. 1.1.2). There are four steps of controlled regulation: namely external, introjected, identified, and integrated regulations. The less autonomous forms of extrinsic motivation are external and introjected regulations. The more autonomous forms of extrinsic motivation encompass identified and integrated regulations (Fig. 1.1.2).

The scientific problem related to the motivation of students for learning Mathematics is posed as a question: what is the motivation of the new generation school students for learning Mathematics?

The object of the research is the motivation of students for learning Mathematics.

The aim of the research is to reveal the peculiarities of motivation of lower secondary school students for learning Mathematics.

The objectives of the research are as follows:

1. What is the extrinsic and intrinsic motivation of lower secondary school students for learning Mathematics?
2. How is the extrinsic motivation of lower secondary school students related to intrinsic motivation?

4.1.2 Method of research

The national research provides data on the student attitudes towards learning Mathematics and their motivation to learn this subject. The secondary analysis of school students' motivation is based on the theoretical model of motivation. The data of the national research (2012) on 8th grade students are analysed according to the model of motivation continuum. One group of questions (M1) concerning the motivation for learning Mathematics is included into the Questionnaire about Mathematics for 8th grade students (Questionnaire about mathematics for 8th grade students, 2012). Exploratory factor analysis (EFA) of questions of group M1 has been conducted.

The research sample of the national research was reliable as it involved 4479 students. The total population was 35,000 8th grade students in 2008 (EMIS – Education Management Information System). The confidence interval being 5 %, confidentiality level is 95 %. Hence, the research sample should have included 379 respondents. Therefore, the probability (confidentiality level) is 95 % that the obtained data can shift only by 5 % from the population parameters (confidence interval).

4.1.3 Results

4.1.3.1 The model of data analysis: exploratory factor analysis of the motivation of 8th grade students for learning Mathematics

Factor analysis is used to investigate variable relationships of complex concepts such as motivation. It allows investigating the concepts that are not easily measured directly by reducing a large number of variables into a few underlying interpretable factors. EFA is employed to investigate the variable of motivation. EFA complies with the research aim, as it is a statistical technique that is used to reduce data to a smaller set of summary variables and to explore the underlining theoretical structure of the phenomena.

The variations in eight observed variables (*You like Maths M1a; You are a gifted mathematician M1b; You like Maths lessons M1c; Learning Maths is interesting M1d; Maths is important for learning other subjects M1e; You learn Maths because it is important in getting entry to a gymnasium or high school M1f; You learn Maths because it is important if you want to get an interesting job M1g; You learn Maths because you want to get a well-paid job M1h; You learn Maths because you need it for everyday life M1i*) in the question group M1 can be reduced to a lower number of unobserved variables called factors. EFA seeks to uncover the underlying structure of a relatively large set of variables. A priori assumption was that any indicator may be associated with any factor.

The key concept of factor analysis is that multiple observed variables have similar patterns of responses because they are all associated with a latent (i.e. not directly measured) variable (Field, 2000). In our case, the observed variables of question group M1 can be associated with the latent variable, i.e. different levels of motivation for learning Maths: amotivation, extrinsic motivation, and intrinsic motivation (Ryan & Deci, 2002).

In this monograph, EFA was carried out employing the factor analysis diagram of Rietveld and Van Hout (1993, p. 291). It offered an overview of the steps in factor analysis: reliable measurements, correlation matrix, factor analysis versus principal component analysis (PCA), the number of factors to be retained, factor rotation, and the use and interpretation of the results (Rietveld & Van Hout, 1993). Applying factor analysis, the fact that variables can be normally distributed and measured at a range level has been taken into account (Field, 2000, p. 444). The skewness (from −1 to +1) and kurtosis (from −1 to +1) of the variables from question group M1 were well within a tolerable range in assuming a normal distribution (Gravetter & Wallnau, 2014; Tab. 4.1.1).

Tab. 4.1.1: Descriptive statistics for the motivation of 8th grade students for learning Mathematics (N = 4479)

Question	No. of items	M (SD)	Skewness	Kurtosis
You like Maths	M1a	2.61 (0.905)	−0.201	−0.730
You are gifted in Maths	M1b	2.44 (0.840)	−0.003	−0.596
You like Maths lessons	M1c	2.55 (0.834)	−0.183	−0.529
Learning Maths is interesting	M1d	2.58 (0.824)	−0.283	−0.447
Maths is important for learning other subjects	M1e	3.06 (0.675)	−0.750	0.954
You learn Maths because it is important in getting entry to a gymnasium or high school	M1f	3.10 (0.678)	−0.628	0.959
You learn Maths because it is important if you want to get an interesting job	M1g	2.84 (0.781)	−0.339	−0.218
You learn Maths because you want to get a well-paid job	M1h	3.04 (0.732)	−0.590	0.434
You learn Maths because you need it for everyday life	M1i	2.97 (0.768)	−0.589	0.275

Factorability was examined by measuring the sampling adequacy. The Kaiser-Meyer-Olkin test (KMO-test) was used for sampling adequacy (KMO-test). The sample is adequate if the value of KMO is greater than 0.5. It was disclosed that KMO = 0.783 for the observed variables of the question group M1.

The starting point of factor analysis is correlation matrix, in which the inter-correlations between the studied variables are presented. The dimensionality of this matrix can be reduced by "looking for variables that correlate highly with a group of other variables but correlate very badly with variables outside of that group" (Field, 2000, p. 424). The variables with high inter-correlations could measure one underlying variable, which is called a 'factor'. Inter-correlation was checked by using Bartlett's test of sphericity (c^2 (36) = 4,792,617, $p < .05$). It means that the data can be used for factor analysis.

Factorability of research data was assured by anti-image correlation. All elements on the diagonal (MSA- Measure of Sampling Adequacy) axis of the anti-image correlation matrix should be greater than 0.5 if the sample is adequate (Field, 2000, p. 446). All variables were suitable for factor analysis (Tab. 4.1.2).

After having obtained the correlation matrix, the PCA was employed to investigate the variable of the question group M1. There are several rules of determining how many factors are appropriate for data analysis: only those factors

Tab. 4.1.2: Anti-image correlation matrix of question group M1 on the motivation for learning Mathematics

	1. You like Maths	**2. You are gifted in Maths**	**3. You like Maths lessons**	**4. Learning Maths is interesting**	**5. Maths is important for learning other subjects**	**6. You learn Maths because it is important in getting entry to a gymnasium or high school**	**7. You learn Maths because it is important if you want to get an interesting job**	**8. You learn Maths because you want to get a well-paid job**	**9. You learn Maths because you need it for everyday life**
1	0.851[a]	0.345	0.446	0.246	0.017	0.018	0.051	0.022	0.042
2		0.921[a]	0.083	0.098	0.077	0.035	0.012	0.010	0.017
3			0.845[a]	0.441	0.072	0.027	0.020	0.024	0.004
4				0.884[a]	0.059	0.005	0.034	0.027	0.085
5					0.906[a]	0.306	0.170	0.098	0.181
6						0.880[a]	0.156	0.262	0.027
7							0.876[a]	0.419	0.045
8								0.821[a]	0.267
9									0.906[a]

[a] Measures of Sampling Adequacy (MSA)

Tab. 4.1.3: Initial eigenvalues of factors and rotation sums of squared loadings of questions in group M1

Component	Initial Eigenvalues			Extraction Sums of Squared Loadings			Rotation Sums of Squared Loadings		
	Total	% of Variance	Cumulative %	Total	% of Variance	Cumulative %	Total	% of Variance	Cumulative %
1	4.519	50.206	50.206	4.519	50.206	50.206	3.186	35.404	35.404
2	1.712	19.017	69.222	1.712	19.017	69.222	3.044	33.818	69.222
3	0.678	7.529	76.752						
4	0.512	5.690	82.442						
5	0.440	4.890	87.331						
6	0.409	4.543	91.875						
7	0.317	3.525	95.399						
8	0.235	2.609	98.008						
9	0.179	1.992	100.000						

Extraction Method: PCA.

with an eigenvalue larger than 1 (Guttman-Kaiser rule) should be retained; as well as the factors which, in total, account for about 70–80 % of the variance should be kept (Field, 2000, pp. 436–437; Rietveld & Van Hout, 1993, pp. 273–274). Initial eigenvalues indicated that the first three factors explained 38 %, 16 %, and 14 % of the variance respectively (Tab. 4.1.2). Hence, they explained 78 % of variance. The fourth and all the following factors explained smaller and smaller portions of the variance. Two factors (Tab. 4.1.3) corresponded to Guttman-Kaiser's rule because their eigenvalues > 1 explained 69 % of variance. It means that two factors were appropriate for data analysis.

The rotation method *Varimax* was used to simplify the factor interpretations in PCA. The goal of rotation is to enhance the link between the variables of one or another factor, i.e. to create their clusters by greater factorial load. After rotation, the dispersion of the first factor decreased to 35 %, whereas the second factor increased almost to 34 % (Rotation Sums of Squared Loadings), (Tab. 4.1.3).

The last step in EFA was the interpretation of the results (Rietveld & Van Hout, 1993, p. 291) based on factor loadings, which are presented in the rotated component matrix (Tab. 4.1.4).

The relationship of each variable to the underlying factor is expressed by factor loading. One could say that the variable *You like Maths* has a correlation of 0.900 with Factor 1. This would be considered a strong association in factor analysis. Three other variables, *You like Maths lessons*, *Learning Maths is interesting*, *You are gifted in Maths*, are also associated with Factor 1. Bearing in mind

Tab. 4.1.4: Rotated Component Matrix of observed variables of the motivation for learning Mathematics and the levels of motivation.

Observed variables	Latent variables: factors		Levels of motivation
	F1	F2	
You like Maths	0.900		**Intrinsic motivation** Interest. engagement. satisfaction
You like Maths lessons	0.890		
Learning Maths is interesting	0.864		
You are gifted in Maths	0.788		
You learn Maths because you want to get a well-paid job		0.855	**Extrinsic motivation** Identified regulation: personal relevance. conscious evaluation
You learn Maths because it is important if you want to get an interesting job		0.792	
You learn Maths because it is important in getting entry to a gymnasium or high school		0.769	
Maths is important for learning other subjects		0.728	
You learn Maths because you need it for everyday life		0.632	

Extraction Method: PCA.
Rotation Method: Varimax with Kaiser Normalization.
[a] Rotation converged in 3 iterations.

that the loading of variables highly depended on Factor 1, we referred to it as intrinsic motivation: interest, engagement, and satisfaction.

The variables *You learn Maths because you want to get a well-paid job*, *You learn Maths because it is important if you want to get an interesting job*, *You learn Maths because it is important in getting entry to a gymnasium or high school*, had high factor loadings on Factor 2. They seemed to indicate the identified regulation: personal relevance and conscious evaluation, so we entitled Factor 2 as extrinsic motivation (identified regulation: personal relevance and conscious evaluation). The variables *Maths is important for learning other subjects*, *You learn Maths because you need it for everyday life* also had a correlation with Factor 2. These variables indicated integrated regulation: awareness and synthesis with oneself, and belonged to the factor of extrinsic motivation.

The data of factor analysis of questions in group M1 revealed two levels of motivation for learning: extrinsic motivation and intrinsic motivation. The analysis of the questionnaire of the national research (2012) on 8th grade students (questionnaire about mathematics for 8th grade students, 2012) highlighted a question related to extrinsic motivation: *It is important for you to know what and*

Tab. 4.1.5: Theoretical model of data analysis of motivation for learning

Observed variables	Type of regulation	Levels of motivation
You like Maths (M1a)	Interest. Engagement. satisfaction	**Intrinsic motivation**
You like Maths lessons (M1c)		
Learning Maths is interesting (M1d)		
You are gifted in Maths (M1b)		
You learn Maths because you want to get a well-paid job (M1h)	Identified regulation: personal relevance. conscious evaluation	**Extrinsic motivation**
You learn Maths because it is important if you want to get an interesting job (M1g)		
You learn Maths because it is important in getting entry to a gymnasium or high school (M1f)		
Maths is important for learning other subjects (M1e)		
You learn Maths because you need it for everyday life (M1i)		
It is important for you to know what and why you are studying (M7a)		

why you are studying (M7a). This question was included in the theoretical model of analysis of the motivation for learning (Tab. 4.1.5).

4.1.3.2 Extrinsic motivation of students for learning Mathematics

The main idea of SDT is that students are growth-oriented by fulfilling the basic psychological needs for autonomy, competency, and social relatedness. Social environments can influence these psychological needs. They are assumed to be innate for all humans despite their cultural differences and in all situations; they do not affect behaviour in a vacuum but their effects are mediated by social-cognitive constructs that are more context-specific, such as the regulatory style or motivation orientation of a person (Ryan & Deci, 2000). Students have different social backgrounds with different prior experiences related to learning situations. A person's motivation in a particular situation is a function of the immediate social context and of the person's inner resources that, have been affected by the prior interactions with social contexts (Ryan & Deci, 2002; Vansteenkiste, Williams, & Resnicow, 2012).

It is unrealistic to imagine that the whole learning can be intrinsically motivated, because it is not possible to make all the goals of the curriculum intrinsically motivating, and in a classroom situation it is not always possible to give students choices about what they should learn (Byman & Kansanen, 2008). Loukomies (2013) explains that "Keeping in mind the wide variety of students' previous experiences concerning the fulfilment of basic psychological needs and the variety of environments from which students come, it is easy to grasp that the science class environment is not experienced identically or even similarly by all the students; they do not have comparable prior expectations and emotions towards an activity or topic" (Loukomies, 2013). According to Loukomies (2013), educational situation also influences motivation for learning. It means that the level of motivation for learning corresponds to four vectors: a vector of psychological needs, a vector of students' experience, a vector of social background and a vector of educational situation. The sum of these vectors determines the level of motivation: amotivation, extrinsic motivation, and intrinsic motivation.

The types of extrinsic motivation are differentiated by the degree of autonomy in the regulation of behaviour. Different stages reflect the degree to which the value of behaviour is internalised (Ryan & Deci, 2000). The data of the national research (2012) on 8th grade students allows conducting analysis of extrinsic motivation for learning Mathematics and Lithuanian language at the level of the identification of extrinsic motivation (Data of the national research, 2012). Identification involves a conscious acceptance of behaviour as being important in order to achieve personally valued outcomes. At the stage of identified regulation, students see the benefits of activity. With *identified* regulation, action begins to be integrated within the students' sense of self: *You learn Maths because you want to get a well-paid job (M1h), because it is important to get an interesting job (M1g), because it is important to get entry to a gymnasium or high school (M1f), Maths is important for learning other subjects (M1e), Maths is necessary for everyday life (M1i).* The mentioned questions about the benefits of learning Maths can be sorted into two groups. The first group of questions is related to social benefits: *You learn Maths because you want to get a well-paid job (M1h), because it is important to get an interesting job (M1g).* The second group of questions is related to educational benefits: *Maths is important for learning other subjects (M1e), You learn Maths because it is important in getting entry to a gymnasium or high school (M1f).*

The Friedman test for the analysis of extrinsic motivation was used. It is a nonparametric test that compares three or more matched or paired groups (Tab. 4.1.6). Friedman test could be employed in our research because our data on extrinsic motivation for learning Maths met the following four assumptions: a group that is measured on three or more different occasions, the group is a random sample

Tab. 4.1.6: Extrinsic motivation of school students for learning Mathematics: Friedman's test mean ranks

Questions about extrinsic motivation: identified level	**Benefits for students**	**Mean Rank**
Maths is important for learning other subjects (M1e)	Educational	3.10
You learn Maths because it is important in getting entry to a gymnasium or high school (M1f)	Educational	3.19
You learn Maths because it is important if you want to get an interesting job (M1g).	Social	2.69
You learn Maths because you want to get a well-paid job (M1h).	Social	3.08
You learn Maths because you need it for everyday life (M1i)	Social	2.94

from the population, dependent variables are measured at the ordinal level, and the samples are normally distributed.

A higher sum of ranks is associated with higher ranks (Tab. 4.1.6). The analysis of questions *Maths is important for learning other subjects (M1e); You learn Maths because it is important in getting entry to a gymnasium or high school (M1f)* is associated with higher ranks. It means that educational benefits of learning Maths are more important than social benefits (Tab. 4.1.6). The data of Freedman test (c^2 (4) = 153,627, $p < .05$) show a statistically significant difference between the answers to different questions on extrinsic motivation for learning Mathematics at the identified level. The p-value for the data is less than the significance level of 0.05. The null hypothesis is rejected and it is concluded that at least two of five types of questions related to the identified level of extrinsic motivation for learning Mathematics have a different effect. These results indicate that educational benefits of learning Mathematics are more effective for students than social benefits.

4.1.3.3 Intrinsic motivation of school students for learning Mathematics

Intrinsic motivation is characterised by self-determined behaviour. Students having self-determined behaviour receive higher grades, learn better, are more satisfied, and have more positive emotions towards school in general (Deci, Ryan, & Guay, 2013; Guay, Lessard, & Dubois, 2016). The factor analysis (Tab. 4.1.5) of the data of the national research (2012) on 8th grade students reveals that there are four questions related to intrinsic motivation for learning Mathematics: *You*

Tab. 4.1.7: Intrinsic motivation of Students for Learning Mathematics: valid percentages

Questions	**Answers rankings**			
	Strongly disagree	**I do not agree**	**Accept**	**Totally agree**
You like Maths	13.1	29.1	42.2	15.6
You are gifted in Maths	13.6	38.6	38.5	9.3
You like Maths lessons	11.3	33.4	44.3	11.0
Learning Maths is interesting	11.1	30.7	47.9	10.3

like Maths (M1a); You like Maths lessons (M1c); Learning Maths is interesting (M1d); You are gifted in Maths (M1b).

Intrinsically motivated students get engaged in certain activities freely and with enjoyment. According to Ryan and Deci, "the basis of intrinsic motivation is interest" (Ryan & Deci, 2009, p. 177). There are four phases of interest development (Hidi & Renninger, 2006; Hidi, 2011): triggered situational interest, maintained situational interest, emerging personal interest, and well-developed personal interest. According to Loucomies (2013), in a classroom triggered situational interest is sparked by environmental features, is mainly supported externally, and may cause positive changes at both cognitive and emotional levels. Meanwhile, maintained situational interest is related to students' continuous, externally supported involvement in educational situation. Maintained situational interest may transform into emerging personal interest. Students have a relatively enduring predisposition for being engaged in the activities in the fourth phase. Regrettably, the questionnaire of the national research (2012) of 8th grade students does not allow revealing the phase of interest development for learning Maths (questionnaire about mathematics for 8th grade students, 2012).

The descriptive statistical analysis of the questions on intrinsic motivation reveals that students are intrinsically motivated for learning Maths (Tab. 4.1.7). A valid percentage of students' answers to different questions on intrinsic motivation for learning Mathematics shows that students like Maths, and Maths lessons, as well as have interest for learning Maths (Tab. 4.1.7). Interest is considered as "a unique motivational variable, as well as a psychological state that occurs during interactions between persons and their objects of interest, and is characterised by increased attention, concentration and affect" (Hidi, 2006, p. 70). Our study shows that 58.2 % of the respondents have a positive attitude to the subject of Mathematics: 47.9 % agree and 10.3 % totally agree that learning Maths is interesting.

Tab. 4.1.8: Intrinsic motivation of school students for learning Mathematics: Freedman's test means ranks

Questions on intrinsic motivation	Mean Rank
You like Maths	2.63
You are gifted in Maths	2.31
You like Maths lessons	2.51
Learning Maths is interesting	2.55

Tab. 4.1.9: Correlation of variables of intrinsic motivation for learning Mathematics: Spearman correlation coefficients

	You like Maths	You are gifted in Maths	You like Maths lessons	Learning Maths is interesting
You like Maths	1.000	0.670**	0.790**	0.744**
You are gifted in Maths		1.000	0.586**	0.561**
You like Maths lessons			1.000	0.767**
Learning Maths is interesting				1.000

** Correlation is significant at the 0.01 level (2-tailed).

Friedman test was used for the analysis of intrinsic motivation questions (Tab. 4.1.8). The data of intrinsic motivation were suitable for Friedman test because one group was measured on four occasions, whereas the dependent variable was measured at the ordinal level.

The data of Freedman test (c^2 (3) = 89,239, $p < .05$) show a statistically significant difference. When Friedman test is positive (p less than the selected significance level), a table (Tab. 4.1.8) is displayed showing which of the variables are significantly different from other variables. The variable (*You like Maths*) is significantly different from variables (2), (3), and (4), which correspond to *You are gifted in Maths, You like Maths lessons, Learning Maths is interesting.* The variable *You like Maths* has an emotional background related to the engagement of students towards Mathematics. The variable *You are gifted in Maths* is associated with self-confidence about mathematical capacities of students, *You like Maths lessons* - with educational environment in classroom, whereas *Learning Maths is interesting* - with interest of students. The data analysis demonstrates (Tab. 4.1.8) that the variable based on engagement *You like Maths* has a higher mean rank (2.63).

Aiming at determining the relations between the variables of intrinsic motivation, a correlation analysis was completed (Tab. 4.1.9). The strongest statistically significant correlation was determined between the students' intrinsic motivation variable *You like Maths* and the variable *You like Maths lessons* (r = 0.790**, p = 0.01). The weakest yet statistically strong correlation was determined between the students' intrinsic motivation variable *You like Maths* and the variable *You are gifted in Maths* (r = 0.670**, p = 0.01). It means that the variable concerning educational environment in the classroom *You like Maths lessons* is more related to the intrinsic motivation variable *You like Maths* than to the variable of self-confidence *You are gifted in Maths.*

4.1.3.4 Links between intrinsic and extrinsic motivation for learning Mathematics

Intrinsic motivation is an important type of motivation, but often students are not intrinsically motivated. The motivation challenge that teachers encounter in their educational practice is to find ways from extrinsic to intrinsic motivation. The main difference between intrinsic and extrinsic motivation is that the former refers to doing something because it is inherently interesting or enjoyable, and the latter refers to doing something because it leads to a separable outcome (Ryan & Deci, 2000).

There is a close relationship between extrinsic and intrinsic motivation, which is explained by the OIT. According to OIT, the regulation of behaviour may be autonomous (self-determined) or controlled depending on the degree of internalisation. Intrinsically motivated individuals get freely engaged in certain activities and are led by the feelings of interest and enjoyment (Ryan & Deci, 2002). Intrinsic motivation arises from a desire to learn a topic due to its inherent interests, self-fulfilment, enjoyment, and achievement of mastery of the subject (Ryan & Deci, 2009). Intrinsic motivation is very important at school as "schools are not day camps or recreational centres" (Brophy, 2004, p. 12). Intrinsic motivation contributes to achievement, especially when people are self-employed. This actuality poses a question – how to encourage intrinsic motivation?

The data of the national research (2012) on 8th grade students allows carrying out a correlation analysis of the variables of intrinsic and extrinsic motivation at the identified level. The identified level is a form of extrinsic motivation, which is based on regulation through identification. Here, a person identifies the personal importance of behaviour, and thus, accepts its regulation as his or her own. (Rayan & Deci, 2000). Friedman's test on extrinsic motivation variables reveals that educational importance of learning Mathematics for school students is more effective than social importance. It is relevant to explore how the variables of intrinsic motivation of students for learning Mathematics are related to the educational and the social variables of extrinsic motivation.

Tab. 4.1.10: Statistical relations of intrinsic and extrinsic motivation variables: Spearman correlations coefficients

	Intrinsic motivation				Extrinsic motivation: identified level				
	1	**2**	**3**	**4**	**5**	**6**	**7**	**8**	**9**
1[a]	1.000	0.670**	0.790**	0.744**	0.360**	0.272**	0.323**	0.228**	0.295**
2		1.000	0.586**	0.561**	0.343**	0.268**	0.289**	0.230**	0.245**
3			1.000	0.767**	0.374**	0.279**	0.318**	0.224**	0.295**
4				1.000	0.368**	0.274**	0.318**	0.230**	0.301**
5					1.000	0.594**	0.536**	0.526**	0.451**
6						1.000	0.521**	0.600**	0.386**
7							1.000	0.640**	0.410**
8								1.000	0.497**
9									1.000

** Correlation is significant at the 0.01 level (2-tailed)
[a] The names of variables:
1 You like Maths
2 You are a gifted in Maths
3 You like Maths lessons
4 Learning Maths is interesting
5 Maths is important for learning other subjects
6 You learn Maths because it is important in getting entry to a gymnasium or high school
7 You learn Maths because it is important if you want to get an interesting job
8 You learn Maths because you want to get a well-paid job
9 You learn Maths because you need it for everyday life

Spearman correlation analysis of intrinsic and extrinsic motivation variables (Tab. 4.1.10) shows a statistically significant positive correlation between all the variables of intrinsic and extrinsic motivation. The magnitude of the correlation coefficient indicates the strength of the association. In our case, the association is weak (0.1–0.3), moderate (0.3–0.5), and strong (over 0.5). All the correlations from both groups are statistically significant at the level $p = 0.01$.

Weak and moderate correlations were determined between intrinsic motivation variables and extrinsic motivation variables. However, strong correlations were detected within the variables of intrinsic motivation. Strong correlations were also established within the variables of extrinsic motivation (Tab. 4.1.10). The correlations within the variables of intrinsic motivation for learning Mathematics were stronger than within the variables of extrinsic motivation.

An analysis of extrinsic motivation variables of the educational identification group (*Maths is important for learning other subjects; You learn Maths because it is important in getting entry to a gymnasium or high school*) and social identification group *(You learn Maths because it is important if you want to get an interesting job; You learn Maths because you want to get a well-paid job; You learn Maths because you need it for everyday life)* was completed. Statistically moderate correlations were determined between the variable *Maths is important for learning other subjects* and all the variables of intrinsic motivation: *You like Maths* ($r = 0.360^{**}$; $p = 0.01$); *You are gifted in Maths* ($r = 0.343^{**}$; $p = 0.01$); *You like Maths lessons* ($r = 0.374^{**}$; $p = 0.01$); *Learning Maths is interesting* ($r = 0.368^{**}$; $p = 0.01$). These correlation coefficients were stronger than the ones of intrinsic motivation variables and variables of social identification group: *You learn Maths because it is important in getting entry to a gymnasium or high school; You learn Maths because it is important if you want to get an interesting job; You learn Maths because you want to get a well-paid job; You learn Maths because you need it for everyday life* (Tab. 4.1.10).

This data suggests that educational identification is important for the promotion of intrinsic motivation of school students for learning Mathematics. The internalised role of Mathematics for learning other subjects engages students for learning Mathematics (*You like Maths; You like Maths lessons; Learning Maths is interesting)* and increases their self-confidence (*You are gifted in Maths).* It is important to organise Maths lessons so that they promote educational identification for learning Mathematics. Consequently, it is important to organise Maths lessons in the way that promotes the internalised role of Mathematics for learning others subject. The analysis of school students' motivation also highlights the engaging role of cross-curricular education in Maths education at school.

4.1.4 Conclusions

Motivation can be seen as a result of interactions between an individual's need system and the environmental factors. The sub-theories of SDT - OIT describe motivation as a continuum: non-regulation, extrinsic regulation, and intrinsic regulation. Types of extrinsic motivation are differentiated by the degree of autonomy in the regulation of behaviour. There are four steps of controlled regulation: external, introjected, identified, and integrated regulations. The third step of extrinsic motivation for learning Mathematics involves a conscious acceptance of the school students' behaviour as being important in order to achieve personally valued outcomes: *You learn Maths because you want to get a well-paid job, because it is important to get an interesting job, because it is important in getting entry to a gymnasium or high school, Maths is important for learning other subjects, Maths is necessary for everyday life.* The aforesaid questions about the benefits of learning Maths can be sorted into two groups. The benefit of learning Mathematics for students can be social: *You learn Maths because you want to get a well-paid job, because it is important to get an interesting job, Maths is necessary for everyday life*, or educational: *Maths is important for learning other subjects*, and *because it is important in getting entry to a gymnasium or high school.* The secondary school students perceive the educational benefit for learning Mathematics more. It means that the educational benefit of learning Mathematics for students is more important than the social benefit. The data of Friedman test (c^2 (4) = 153,627, $p < .05$) confirms a statistically significant difference between the students' answers to different questions about the educational and social benefits of learning Mathematics.

Intrinsic motivation is characterised by self-determined behaviour. Intrinsically motivated students get engaged in certain activities freely and with enjoyment. Four variables of the students' intrinsic motivation of were analysed: *You like Maths; You are gifted in Maths; You like Maths lessons; Learning Maths is interesting.* The variable *You like Maths* has an emotional background related to the engagement of students in Mathematics. The variable *You are gifted in Maths* is associated with the students' self-confidence about their mathematical capacities, *You like Maths lessons* - with educational environment in the classroom, and Learning *Maths is interesting* - with the interests of students. Friedman test of the analysis of intrinsic motivation revealed (c^2 (3) = 89,239, $p < .05$) a statistically significant difference between these variables. According to Friedman test, the variable based on engagement *You like Maths* has a higher mean rank (2.63). The correlation analysis discloses the strongest correlation between this variable and the variable of the educational environment in the classroom *You like Maths lessons* (r = 0.790**, p = 0.01). These data confirm the positive role of educational environment in the classroom in relation to the engagement for learning Mathematics.

According to OIT, learner motivation develops from amotivation to extrinsic and intrinsic motivation. The correlations analysis of extrinsic motivation variables of the educational identification group (*Maths is important for learning other subjects; You learn Maths because it is important in getting entry to a gymnasium or high school*) and social identification group *(You learn Maths because it is important if you want to get an interesting job; You learn Maths because you want to get a well-paid job; You learn Maths because you need it for everyday life)* reveals that educational identification is more related to intrinsic motivation of school students for learning Mathematics.

The internalised role of Mathematics for learning other subjects (*Maths is important for learning other subjects*) gets the students more engaged in learning Mathematics than the internalised role of Mathematics in getting entry to a gymnasium or high school (*You learn Maths because it is important in getting *entry to a gymnasium or high school*). Consequently, it is important to organise Maths lessons in the way that promotes the internalised role of Mathematics for learning other subjects. The analysis of school students' motivations reveals the engaging role of cross-curricular Mathematics education at school.

4.2 Motivation of school students for learning Mathematics and autonomy Support

By Palmira Pečiuliauskienė

The section deals with the role of perceived support for autonomous motivation in school, and autonomous motivation for homework out of school. The trans-contextual model has been adapted to examine the effects of the perceived autonomy support as well as relation between autonomous motivation of school students for Mathematics activities in the classroom and their autonomous motivation for doing Mathematics homework outside school.

4.2.1 Introduction

People need to feel competent, autonomous, and affiliated with others when doing a task (Shumow, Schmidt, & Zaleski, 2013). A theme running through the SDT theoretical line on autonomy in education is a new approach to the activity of students and teachers with a shift towards teachers as providers of guidance and scaffolding for student learning, as well as with a shift towards students awarded more autonomy over their own learning process. Teachers have to be "both autonomy supportive and involved, natural states of intrinsic motivation are less likely to be undermined, and the internalisation and integration of extrinsically motivated behaviours will be facilitated" (Deci & Ryan, 2000, p. 238). "Such instructional behaviours as, for instance, listening to students, creating time for independent work, giving student opportunity to talk, and being responsive to students' questions and experience correlated positively with students' perceptions of autonomy" (Byman, Lavonen, Juuti, & Meisalo, 2012).

According to Stefanou et al. (2004), autonomy support can be procedural (students are allowed to choose and handle their own materials, take an active part in hands-on activities), cognitive (students may find multiple solutions to problems, receive considerable support in re-evaluating their errors) and organizational (students may make decisions about the layout of the classroom activities). In classrooms with high autonomy support, where students are asked to make cognitive choices related to solution strategies, students show much more enthusiasm and engagement (Stefanou et al., 2004).

Researchers highlight the problem of the ratio between procedural autonomy support and cognitive autonomy support. "Moreover, the differentiation of procedural autonomy support versus cognitive autonomy support offers a possible explanation for why reform-oriented learning settings sometimes do not reach their potential, mainly because they may place too much weight on the provision of procedural autonomy while neglecting the important aspect of cognitive autonomy support. It has been argued that students might be hindered in their learning if given too much freedom with materials" (Furtak & Kunter, 2012, p. 286).

Autonomy support techniques promote autonomous motivation of students in educational practice. Students who perceive their teachers to be autonomy-supportive are more likely to report autonomous motivation and adaptive educational outcomes (Jang, Kim, & Reeve, 2012). According to Hagger et al. (2015), autonomous motivation reflects behaviours out of the sense of choice and ownership. Autonomously-motivated students pay more attention, invest more effort in class and demonstrate better overall grades (Deci et al., 1996, Wong et al., 2002, Reeve & Jang, 2006). Research has shown that teachers, who allow students to adopt a questioning approach, provide encouragement and positive task-related feedback and avoid control, promote greater autonomous motivation in their students (Reeve et al., 1999; Chatzisarantis et al., 2007). Researchers examine the process by which autonomous motivation in the classroom leads to autonomous motivation for educational activities outside formal educational contexts, as well as autonomous motivation for activities in an educational context was strongly associated with autonomous motivation for the related activities outside of the school context (Hagger & Chatzisarantis, 2012; Hagger et al., 2015).

The analysis of autonomy support activity was completed on the *trans-contextual model* which has considerable potential to guide research on the transfer of motivation across classroom and out-of-school contexts (Hagger et al., 2015): perceived support for autonomous motivation in school, autonomous motivation for school activities, autonomous motivation for homework out of school, beliefs about future homework, intentions to do homework in future, homework behaviour, overall grades. Hagger et al. (2015) analysed autonomous motivation at the level of identified regulation and intrinsic motivation. The simplified model of Hagger et al. (2015) (Fig. 4.2.1) was related to the questions on motivation for learning Mathematics at school (Data of the national research, 2012).

The purpose of the current research is to adopt the trans-contextual model (Fig. 4.2.1) to examine the effects of school students' perceived autonomy support and autonomous motivation for Mathematics activities in the classroom

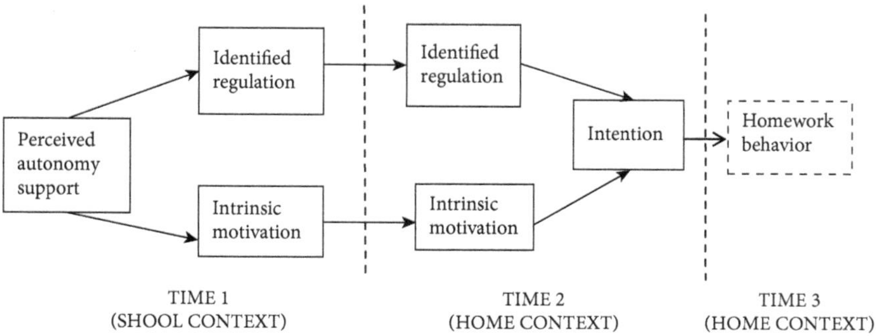

Fig. 4.2.1: The trans-contextual model of perceived support for autonomous motivation: autonomy support at school, intention, homework behavior (simplified model of Hagger et al., 2015)

on their autonomous motivation for Mathematics homework outside school, as well as to adapt this model to analyse issues related to motivation for learning Mathematics.

4.2.2 Method of research

The data of the national research (2012) on 8th grade students allows disclosing the information on the autonomy-supporting learning of Mathematics in the school educational context and outside it (Data of the national research, 2012). In the classroom, a teacher of Mathematics proposes several options for homework, and students can choose which homework assignment to complete (M10d); supports and encourages students' self-confidence (M10e); gives additional and more complex tasks for gifted students (M10f); as well as provides all the necessary assistance (M10g). A student outside the classroom completes the homework assignment on Mathematics independently (M6a); other people can help with the homework (M6b); it is often unclear how to do the homework assignment (M6c); the student tries to do homework as well as possible (M6f); in the classroom, the student discusses with the teacher what difficulties he/she may encounter when doing homework and how to overcome them (M6g). The aforesaid issues enable to analyse autonomy-supportive activities in terms of procedural and organisational autonomy in the school educational context and beyond it.

4.2.3 Results

The Spearman correlation analysis was employed for the analysis of the variable of intrinsic motivation, namely Learning *Maths is interesting* and the supporting

Tab. 4.2.1: Spearman correlation analysis of students' intrinsic motivation for learning Mathematics and autonomy-supportive behaviour in the classroom

	Learning Maths is interesting (1)	**The teacher proposes different tasks (2)**	**The teacher encourages self-confidence (3)**	**The teacher gives additional tasks (4)**	**The teacher provides all the necessary assistance (5)**
(1)	1.000	0.181**	0.338**	0.156**	0.334**
(2)		1.000	0.439**	0.231**	0.297**
(3)			1.000	0.361**	0.564**
(4)				1.000	0.382**
(5)					1.000

autonomous motivation variables in the classroom (Tab. 4.2.1). The data of the correlation analysis revealed weak (0.1–0.3) and moderate (0.3–0.5) correlations. All correlations were statistically significant at the level $p = 0.01$. Two ways of the autonomy-supportive teaching were analysed. First, the teacher's activity involved setting a Mathematics homework assignment for students (*The teacher proposes different homework assignments for students to choose from; The teacher gives additional tasks for gifted students*). Second, autonomous motivation-supporting behaviour was based on interpersonal communication with students in the classroom (*The teacher encourages self-confidence; The teacher provides all the necessary assistance*).

The correlation analysis revealed that the autonomous motivation-supporting activity of the teacher was based on the interpersonal communication with students in the classroom, as well as it was more associated with the intrinsic motivation for learning Mathematics (*Learning Maths is interesting*). Both correlations (*The teacher encourages self-confidence; The teacher provides all the necessary assistance*) were moderate. Meanwhile, the teacher's autonomy-supportive behaviour based on assigning different Maths homework tasks had a weak correlation with the students' intrinsic motivation for learning Mathematics (Tab. 4.2.1).

Spearman correlation analysis of the students' intrinsic motivation for learning Mathematics and autonomy-supportive behaviour doing homework was accomplished (Tab. 4.2.2). There were five questions in the national research (2012) questionnaire on autonomy-supportive behaviour while doing Maths homework: *You do your Maths homework independently (M6a); The family or other people help you to do your Maths homework (M6b); It is often unclear how*

Tab. 4.2.2: Spearman correlation analysis of student's intrinsic motivation for learning Mathematics and autonomy-supportive behaviour at home context

	Learning Maths is interesting (1)	You do your Maths homework independently (2)	Other people help you to do your Maths homework (3)	It is often unclear how to do homework (4)	You try to do your Maths homework as well as possible (5)	You discuss with the teacher in the classroom what difficulties you may encounter (6)
(1)	1.000	0.219**	0.002	−0.323**	0.247**	0.247**
(2)		1.000	0.219**	0.002	−0.323**	0.247**
(3)			1.000	−0.243**	−0.245**	0.307**
(4)				1.000	0.189**	0.007
(5)					1.000	−0.116**
(6)						1.000

** Correlation is significant at the 0.01 level (2-tailed).

to do homework (M6c); You try to do your Maths homework as well as possible (M6f); You discuss with the teacher in the classroom what difficulties you may encounter while doing homework and how to overcome them (M6g).

The data of correlation analysis of the students' intrinsic motivation for learning Mathematics and autonomy-supportive activity while doing homework revealed weak (0.1–0.3) and moderate (0.3–0.5) correlations. The intrinsic motivation for learning Mathematics was not affected by the help of others in doing homework ($r = 0.02$). The correlation between the variable of intrinsic motivation *Learning Maths is interesting* and the variables of autonomy-supportive behaviour while doing homework was weak, yet statistically significant: *You do your Maths homework independently* ($r = 0.219^{**}$, $p = 0.01$); *You try to do your Maths homework as well as possible* ($r = 0.247^{**}$, $p = 0.01$); *You discuss the difficulties with the teacher in the classroom* ($r = 0.257^{**}$, $p = 0.01$).

According to the trans-contextual model of perceived support for autonomous motivation in school (Hagger et al., 2015), *Intention* (Time 2) and *Homework behaviour* (Time 3) were influenced by the identification level of extrinsic motivation. The factor analysis revealed two sub-levels of identification: the educational level and the social level (4.1.5). Hence, the correlation analysis of extrinsic motivation and autonomous motivation-supportive behaviour in the classroom was accomplished (Tab. 4.2.3).

Tab. 4.2.3: Spearman correlation analysis of students' extrinsic motivation for learning Mathematics and autonomy-supportive behaviour in the classroom

	Variables of extrinsic motivation				Variables of autonomy-supportive activity			
	1[a]	2	3	4	5	6	7	8
1	1.000	0.594**	0.536**	0.526**	0.022	0.241**	0.192**	0.252**
2		1.000	0.521**	0.600**	0.009	0.176**	0.202**	0.174**
3			1.000	0.640**	0.112**	0.222**	0.141**	0.237**
4				1.000	0.082*	0.255**	0.174**	0.227**
5					1.000	0.439**	0.231**	0.297**
6						1.000	0.361**	0.564**
7							1.000	0.382**
8								1.000

** Correlation is significant at the 0.01 level (2-tailed).
* Correlation is significant at the 0.05 level (2-tailed).
[a] 1. Maths is important for learning other subjects
2. You learn Maths because it is important in getting entry to a gymnasium or high school
3. You learn Maths because it is important if you want to get an interesting job
4. You learn Maths because you want to get a well-paid job
5. The teacher proposes different tasks for homework
6. The teacher encourages self-confidence
7. The teacher gives additional tasks for gifted students
8. The teacher provides all the necessary assistance

The inter-correlation matrices (Tab. 4.2.3) were divided into two areas: correlations between identified level variables (1–4), and correlations between autonomy-supportive variables (5–8). The correlation between the identified level variables of extrinsic motivation (1–4) and autonomy-supportive variables (5–8) was weak (Tab. 4.2.3). However, seven weak, yet statistically significant correlations were found *(r)* from 0.20 to 0.30. The correlation analysis disclosed that autonomy-supportive behaviour concerning interpersonal communication between a teacher and a student in the classroom was more associated with the identified level of extrinsic motivation: *The teacher encourages self-confidence* (r = 0.241**; r = 0.222**; r = 0.255**); *The teacher provides all the necessary assistance* (r = 0.252**; r = 0.237**; r = 0.227**) (Tab. 4.2.3).

According to the trans-contextual model of perceived support for autonomous motivation at school (Hagger et al., 2015), it is important to reveal correlations between autonomous motivation-supportive behaviour while doing homework and the identified regulation variables (Tab. 4.2.4).

Tab. 4.2.4: Spearman correlation analysis of students' extrinsic motivation for learning Mathematics and autonomy-supportive behaviour in the home context

	Extrinsic motivation variables				**Variables of autonomy-supportive activity**			
	1[a]	**2**	**3**	**4**	**5**	**6**	**7**	**8**
1	1.000	0.594**	0.536**	0.526**	0.256**	0.028	0.264**	0.208**
2		1.000	0.521**	0.600**	0.243**	0.011	0.245**	0.138**
3			1.000	0.640**	0.200**	0.009	0.215**	0.154**
4				1.000	0.164**	0.063	0.223**	0.209**
5					1.000	−0.243**	0.307**	0.180**
6						1.000	0.007	0.106**
7							1.000	0.309**
8								1.000

** Correlation is significant at the 0.01 level (2-tailed).
* Correlation is significant at the 0.05 level (2-tailed).
[a] 1. Maths is important for learning other subjects
2. You learn Maths because it is important in getting entry to a gymnasium or high school
3. You learn Maths because it is important if you want to get an interesting job
4. You learn Maths because you want to get a well-paid job
5. Math homework performs you self
6. Other people help you prepare math homework
7. You trying to make math homework the best possible
8. You discuss with the teacher in the classroom about difficulties

The data analysis revealed that the correlation coefficients between the identification variables (1–4) and autonomous motivation-supportive variables (5–8) were weak (Tab. 4.2.4), although a lot of them are statistically significant. All correlations between the autonomous motivation-supportive variable *You try to do your Maths homework as well as possible* and the extrinsic motivation variables were statistically significant.

The purpose of the current research is to adopt the trans-contextual model to examine the effects of school students' perceived autonomy support for their intrinsic motivation for learning Mathematics in the classroom on their intrinsic motivation for doing Maths homework outside the school. The research contributed to the understanding of the extent to which motivation for learning activities in different educational contexts was related to motivation for similar assignments in the homework context. The data revealed motivational transfer across contexts.

The promotion of students' self-confidence in the classroom (*The teacher encourages self-confidence*) statistically correlated with the variables of extrinsic motivation: *Math is important for learning other subjects* (r = 0.241**; p = 0.01); *You learn Maths because it is important if you want to get an interesting job* (r = 0.222**; p = 0.01); *You learn Maths because you want to get a well-paid job* (r = 0.255**; p = 0.01). Self-confidence affects motivation while trying to do Maths homework as well as possible (r = 0.255**; p = 0.01). Motivation to do Maths homework as well as possible in the home context was statistically related to all variables of extrinsic motivation: *Maths is important for learning other subjects* (r = 0.264**; p = 0.01); *You learn Maths because it is important in getting entry to a gymnasium or high school* (r = 0.245**; p = 0.01); *You learn Maths because it is important if you want to get an interesting job* (r = 0.215**; p = 0.01); *You learn Maths because you want to get a well-paid job* (r = 0.223**; p = 0.01). We expect these results can have important implications for educational practice by demonstrating how the fostering of autonomous motivation in the classroom may affect students' behaviour towards learning activities in the homework context.

4.2.4 Conclusions

Autonomy support gets students engaged in learning Mathematics. The analysis of autonomy-supportive behaviour and intrinsic motivation was completed using the *trans-contextual model*, which has considerable potential to guide research on the transfer of motivation across classroom and homework contexts. Two ways of the autonomy-supportive behaviour were analysed. First, the teacher's behaviour involved setting a Mathematics homework assignment for students (*The teacher proposes different homework assignments for students to choose from; The teacher gives additional tasks for gifted students*). Second, autonomous motivation-supporting behaviour was based on interpersonal communication with students in the classroom (*The teacher encourages self-confidence; The teacher provides all the necessary assistance*). The correlation analysis revealed that the autonomous motivation-supporting activity of the teacher was based on the interpersonal communication with students in the classroom, as well as it was more associated with the intrinsic motivation for learning Mathematics (*Learning Maths is interesting*). Both correlations (*The teacher encourages self-confidence; The teacher provides all the necessary assistance*) were moderate. Meanwhile, the teacher's autonomy-supportive behaviour based on assigning different Maths homework tasks had a weak correlation with the students' intrinsic motivation for learning Mathematics.

The correlation analysis revealed that the teacher's individual autonomy-supportive behaviour towards students in the classroom was more associated with their intrinsic motivation for learning Mathematics similar to the home context. The correlation between the variable of intrinsic motivation *Learning Maths is interesting* and the variables of autonomy-supportive behaviour in the home context was weak, yet statistically significant: *You do your Maths homework independently* (r = 0.219**, p = 0.01); *You try to do your Maths homework as well as possible* (r = 0.247**, p = 0.01); *You discuss the difficulties with the teacher in the classroom* (r = 0.257**, p = 0.01).

The *trans-contextual model* was applied for the investigation of the identified level of students' extrinsic motivation and autonomous motivation-supportive behaviour in the classroom and home contexts. The correlation analysis disclosed that autonomy-supportive behaviour based on the interpersonal communication between the teacher and the student in the classroom correlated with the identified level of extrinsic motivation: *The teacher encourages self-confidence* (r = 0.241**, p = 0.01; r = 0.222**, p = 0.01; r = 0.255**, p = 0.01); *The teacher provides all the necessary assistance* (r = 0.252**, p = 0.01; r = 0.237**, p = 0.01; r = 0.227**, p = 0.01).

The trans-contextual model disclosed the role of the self-confidence of students. The promotion of students' self-confidence in the classroom (*The teacher encourages self-confidence*) statistically correlated with the variables of extrinsic motivation: *Maths is important for learning other subjects* (r = 0.241**; p = 0.01); *You learn Maths because it is important if you want to get an interesting job* (r = 0.22**; p = 0.01); *You learn Maths because you want to get a well-paid job* (r = 0.255**; p = 0.01). Self-confidence affects motivation while trying to do Maths homework as well as possible (r = 0.255**; p = 0.01). Motivation to do Maths homework as well as possible in the home context was statistically related to all variables of extrinsic motivation: *Maths is important for learning other subjects* (r = 0.264**; p = 0.01); *You learn Maths because it is important in getting entry to a gymnasium or high school* (r = 0.245**; p = 0.01); *You learn Maths because it is important if you want to get an interesting job* (r = 0.215**; p = 0.01); *You learn Maths because you want to get a well-paid job* (r = 0.223**; p = 0.01).

5 Discussion

By Palmira Pečiuliauskienė

The current research considers the processes activated in using inquiry-based learning. We have shown how inquiry-based learning is useful for the promotion of the motivation of lower secondary school students for learning Physics and Mathematics in digital and in real environments. The conducted comparative analysis of the motivation for learning Physics in terms of the performance of Physics labs in the real and virtual environment has confirmed the assumption of Nedic, Machotka and Nafalski (2003) that there is no simple answer to the dilemma – which laboratory work is best for students.

The correlation analysis of the national (Lithuania) research data on the learning achievements of Generation Y students reveals no significant effect of using computers on the assessment of the subject of Physics or learning motivation. Surveys accomplished by other researchers (Zacharia & Olympiou, 2010; Changeiywo, Wambugu, & Wachanga, 2011) demonstrate that the application of modern virtual environments enhance motivation for learning Physics and raise interest in the subject. Thus, it is assumed that two approaches can be distinguished while analysing the role of information and communication technology in studying Physics: generalised statistical approach (in our case) and a specific approach (specific experiments of using technologies in a specified place and time). However, the analysis of the employment of virtual learning platforms in specific conditions (Ince, Kirbaslar, Yolcu et al., 2014) merely reveals the effect of specific technologies on the learning process and motivation to study Physics.

Our qualitative research demonstrates that the interest of lower secondary school students in Physics is promoted by the laboratory work conducted in the real environment. The students express their interest in the 'hands-on' activity, where they could fiddle and touch the devices and materials. This conclusion approves of the conclusions of other researchers (Swarat, Ortony, & Revelle, 2012) that 'hands-on' activities empowering students' engagement into technology elicit higher interest. More emphasis should be put on the role of activity in constructing interesting learning environments. While conducting virtual laboratory experiments, students are obliged to accomplish only the pre-programmed assignments. A student acts according to someone else's rules and design. This is profoundly different from a student having an original idea to make or do something.

Another finding of this study deals with the promotion of motivation for learning Physics at digital and hands-on Physics labs. The finding can be affected by the inquiry level (confirmation, structured, guided, and open). The current study on digital and hands-on Physics labs has not explored students' motivation at all inquiry levels. The research involves the analysis of the results of Physics labs that are attributed to structured and guided inquiry in terms of the Inquiry Theory (Banchi & Bell, 2008). We acknowledge this limitation and suggest that the results gained from this exploratory research encourage investigations at the level of structured and open inquiry.

Wolf & Fraser (2008) analysed how eighth-form learners explore activity of static electricity and determine that learners in the inquiry classes worked more closely and offered advice and suggestions. According to the sociological characteristics of generations, the research participants in Wolf and Fraser's (2008) study belonged to Generation Y. Hence, social interaction was important for learners of Generation Y while accomplishing hands-on labs in static electricity. Our research also involves students' social interaction. It appears that the respondents attribute the least amount of points to the need for real social interaction while accomplishing digital labs in Physics (Tab. 3.2.1). The component of their real social interaction at digital Physics labs does not statistically correlate with the intrinsic motivation for learning Physics (Tab. 3.2.3).

Loukomies et al. (2013) maintain that, in terms of inquiry strategy, it is important to employ not only students' prior knowledge but also the basic psychological needs they want to fulfil. The current research about promotion of motivation for learning Physics at digital and hands-on Physics labs embodies design of the enhanced intrinsic motivation by supporting the students' innate psychological needs. The conducted research reveals that basic psychological needs (for autonomy, competency, and social relatedness) are significant for the intrinsic motivation of new generation students for studying Physics. The most expressed among them is the need for competency recognition (Tab. 3.2.3; Tab. 3.2.6).

We agree with Loukomies et al. (2013) that "different learners value aspects intended to enhance motivation differently; and therefore, various motivational features make it easier to affect different learners with one sequence" (p. 2536). A well-designed activity that encompasses support for new generation learners' basic psychological needs and especially for support perceived competency is a reasonable way of enhancing new generation students' intrinsic motivation for accomplishing digital and real Physics labs.

Further contribution of this study is the analysis of students' motivation for learning Physics at the levels of structured and guided inquiry. According to SDT, it is important to fulfil three basic psychological needs. Intrinsic motivation is fostered when we "provide people [with the] opportunity to satisfy their basic psychological needs for competency, relatedness and autonomy" (Deci, Vallerand, Pelletier, & Ryan, 1991, p. 329). From this perspective, motivation can be seen as a result of interactions between an individual's need system and the environmental factors that interfere with or support the process of need fulfilment (Loukomies, 2013).

Physics labs foster the needs for competency of school students. Feeling for competency is strongly expressed at guided and structured Physics labs (Tab. 3.3.1; Tab. 3.3.5). Statistically significant difference between the fulfilment of feeling of competency are revealed at different levels of Physics labs: structured and guided inquiry levels. Why so? The answer lies in the different ways of solving a research problem at structured and guided Physics labs. Different ways of solving a research problem require different competencies of students.

In guided Physics labs, students play the role of an 'original researcher' (Exner, 2014, p. 460) because the laboratory manual provides only the problem and procedures, yet the methods of analysis, communication, and conclusions are for the students to design. The original researcher needs to identify information needs that surround their core information need in order to determine what will be needed to devise an original research plan for their final empirical investigation. Original researchers may have more complex information needs; therefore, their potential costs and benefits are more complex, too (Exner, 2014). Sciences especially heavily rely on the process of questioning and articulating an actionable method of inquiry, such as an experimental hypothesis or a set of formalised research questions (Freedman, 2004). Guided Physics labs emphasise the process of articulating an actionable method of inquiry. They require more competency of students. The results of our research reveal that students feel less competent at guided Physics labs ($\overline{x} = 57.31 \pm 12.51$) as compared to at structured Physics labs ($\overline{x} = 63.04 \pm 17.07$). In structured inquiry Physics labs, students often act as information synthesisers. According to Exner (2014, p. 460), 'information synthesisers' ask a question which is intended to be answered through the process of information retrieval. It requires less effort and competency.

According to SDT, it is important to fulfil the need for autonomy in order to foster intrinsic motivation of students. A theme running through SDT theoretical line about autonomy in education is a new perspective on the activity of students and teachers, with a shift towards teachers as providers of guidance and scaffolding for student learning, as well as with a focus on awarding students with more autonomy over their own learning process. Teachers have to be "both autonomy supportive and involved, natural states of intrinsic motivation are less likely to be undermined, and the internalisation and integration of extrinsically motivated behaviours will be facilitated" (Deci & Ryan, 2000,

p. 238). "Such instructional behaviours as, for instance, listening to students, creating time for independent work, giving student opportunity to talk, and being responsive to the students' questions and experience correlated positively with students' perceptions of autonomy" (Byman, Lavonen, Juuti, & Meisalo, 2012, p. 389). The six characteristics that represent the areas in the labs, where students could act independently, are as follows: Problem/Question, Theory/Background, Procedures/Design, Analysis of Results, Communication of Results, and Conclusions.

Stefanou et al. (2004) state that autonomy support can be 'procedural autonomy' (students are allowed to choose and handle their own experimental materials, as well as take an active part in hands-on activities), 'cognitive autonomy' support (students may find multiple solutions to problems and receive considerable support in re-evaluating their errors), and 'organisational support' (students may make decisions about the layout of the classroom activities) (Stefanou et al., 2004, p. 101). Different levels of Physics labs ensure different kinds of autonomy support. In structured and guided Physics labs, autonomy is procedural and cognitive because students are allowed to choose and handle their own experimental materials and find multiple solutions to problems. In guided Physics labs students have more cognitive autonomy; they act as original researchers searching for multiple solutions to problems by choosing methods of Physics labs data analysis, communication, and conclusions. However the data of our research do not reveal a statistically significant difference ($p > 0.05$) between the feeling of autonomy in structured ($\overline{x} = 57.59 \pm 16.22$) and guided ($\overline{x} = 59.74 \pm 14.24$) Physics labs.

The results of the research performed by Stefanou et al. (2004) showed that student engagement was low in the classes with high and low procedural autonomy support (students were allowed to make choices about the structure of the class). By contrast, in the classrooms with high cognitive autonomy support where students were asked to make cognitive choices related to strategies of the solution, students showed much more enthusiasm and engagement (Stefanou et al., 2004). Autonomy-supportive teacher behaviour can be effective in fostering intrinsic motivation in students (Reeve & Jang, 2006; Reeve & Lee, 2014). The data of our research do not reveal a statistically significant correlation between the data of first subscale *interest/enjoyment* and the fourth subscale *perceived choice/autonomy* at the levels of structured and guided inquiry. This result may be influenced by the absence of students' organisational abilities (students could not make decisions about the layout of the classroom activities).

It is important to discuss the role of responsible research and innovation (RRI) and Socio-Scientific issues (SSI) promoting motivation of new generation students for learning science. "The ability to deal with SSI has been recognized as an important goal of science education" (Evagorou et al., p. 402). It is our contention that SSI is a good tool for successful "passing through a long and ghastly kitchen" (Osborne et al., 2003, p. 1074) for new generation learners. According to Owen et al. (2012), a responsible innovation evokes a collective duty of care, first to rethink what we want from innovation because the societal perception and impacts of technology are difficult (impossible) to predict. Acknowledging the power of innovation to shape our collective future, RRI challenges us, first and foremost, to ask what kind of future we want innovation to bring into the world (Owen et al., 2012).

We investigated how SSI activity gets students engaged in science in the science classroom. We also sought to bring to light how engagement is related to relatedness of students. The essential ingredients of motivation are opportunities to choose, control over the pace and nature of learning, and communication (Osborne et al., 2003). The feeling of relatedness at SSI activity is fulfilled through communication when carrying out mini-projects and dialogues of students with scientists.

The questions are the tools of collaboration. Scholars state that students' learning and motivation can be fostered if students receive deep-reasoning questions (Jurik, Gröschner, & Seidel, 2014). Deep-reasoning questions in the science classroom can provide a rich forum for the exploration of disparate viewpoints. Students' competencies appear in the way they evaluate information pertinent to socio-scientific problems and the ethical issues, as well as in the way they find a dilemma solving solution. By engaging students in the discourse on SSI, teachers can challenge their intrinsic motivation.

There is an international push to improve the effectiveness with which scientists communicate. It is acknowledged that the role of communicating science research to a broad range of audiences is the responsibility of a trained science community (Brownell, Price, & Steinman, 2013; Mercer-Mapstone & Kuchel, 2015). SSI requires a dialogue between scientists and students, as well as scientists and teachers. SSI activity is based on collaboration and teamwork approach, which encourages intrinsic motivation and promotes student responsibility for their learning and communication abilities.

Working actively with an interesting topic may enable students to get to know each other better and even to make friends (Loukomies et al., 2013). To make friends, according to Loukomies et al. (2013), means to be at the level I-Thou according to Buber (1970). Factor analysis of our research revealed that I-Thou level at RRI activity was based on trust and closeness of persons: *I felt like I could really trust this person; I feel close to this person*. According to

Okada (2016), effective dialogue between scientists and non-scientists requires two factors: a trusting relationship for mutual exchange of information and know-how for collaborative decision-making processes. The solution to this dilemma problem was offered by the ENGAGE project, stating that the trust between different RRI actors: students, teachers, and scientists, is required. The data of our research showed that the trust of interpersonal communication was statistically related to the Engagement variables *This activity was fun to do* ($r = 0.248^{**}$, $p= 0.01$), *I thought this activity was quite enjoyable* ($r = 0.227^{**}$, $p = 0.01$), and *while I was doing this activity, I was thinking about how much I enjoyed it* ($r = 0.209^{**}$, $p = 0.01$) (Tab. 3.4.2).

It was possible to find combinations where the correlation between Engagement variables and I-You variables was statistically moderate (Tab. 3.4.2). It was surprising that the correlation between I-You variables (*It is likely that this person and I could become friends if we interacted a lot; I'd like a chance to interact with this person more often*) and *engagement* variables was a little bit higher than the correlation between I-Thou and Engagement variables. The moderate correlations between I-You variables and Engagement variables can be explained on the basis of expectancy-value theory.

The phenomenon of motivation for learning Physics can be analysed in two aspects: by the content of Physics and by the language of Physics. As it is already mentioned in the monograph, Mathematics is language of Physics (Redish & Kuo, 2015). The relation between the achievements in Mathematics and Physics are reported by many researchers (Ben, 2010; Olusola & Rotimi, 2012). It is important to discuss the promotion of motivation for learning Physics in the light of the motivation for learning Mathematics. We remember our question: do the psychological needs for the motivation to learn Mathematics and Physics overlap or are they different? We have chosen the SDT as the theoretical background for discussion. According to Ng et al. (2015), students having good motivated strategies for learning Mathematics seem to perceive competency as a more prevalent need than autonomy. We have analysed the relation of autonomy and competency with students' motivation for learning Mathematics.

Our findings on the promotion of students' motivation in digital Physics labs reveal that the strongest and statistically significant correlation is manifested between the students' intrinsic motivation for learning Physics and their feeling of competency ($r = 0.462^{**}$, when $p = 0.01$), as well as between intrinsic motivation for learning Physics and the feeling of autonomy while accomplishing a lab (the first and fifth subscales) ($r = 0.235^{*}$, when $p = 0.01$). The correlation between the students' intrinsic motivation for learning Physics and their feeling of competency is stronger than the correlation between motivation to learn Physics and the feeling of autonomy while accomplishing a digital lab.

Our findings on the promotion of students' motivation in guided inquiry Physics labs confirm that students feel a strong sense of recognition of competency in them. The findings of our research are similar to the findings of Ng et al. (2015) on learning Mathematics, which show that students having good motivated strategies for learning Mathematics seem to perceive competency as a more prevalent need than autonomy. Our findings also prove that autonomy support gets students engaged in learning Mathematics (Ng et al., 2015). We have analysed two ways of autonomy-supportive behaviour. The first way involves teachers' autonomy-supportive behaviour based on giving Maths tasks for students (*The teacher proposes different tasks for homework, from which a student can choose; the teacher gives additional tasks for gifted students*). The second way embraces teachers' autonomy-supportive behaviour based on interpersonal communication with students in the classroom (*The teacher encourages self-confidence; the teacher provides necessary assistance*). The correlation analysis reveals that the teacher's autonomy-supportive behaviour based on interpersonal communication with the students in the classroom is more related to intrinsic motivation for learning Mathematics (*Learning Maths is interesting*).

We have disclosed the role of interpersonal communication concerning motivation for learning science during educational practice based on RRI activity. Our findings confirm that perceived competency and social relatedness are linked to the learning of science. The research reveals weak, yet statistically significant relationship between the engagement of students in science and relatedness (student–teacher–scientist) at RRI activity. We have also revealed that basic psychological needs are different in strength of the enjoyment and keenness for learning science.

Our data indicate a positive effect of inquiry-based learning of Physics and Mathematics on students' intrinsic motivation. Their sense of autonomy for learning, as well as their sense of competency and social relatedness promotes motivation for learning Physics and Mathematics. These results echo the ideas espoused for motivational learning in Deci and Ryan's SDT (Ryan & Deci, 2000; Deci & Ryan, 2012). These results echo the ideas of both Deci and Ryan: Physics content and by Physics language.

6 Conclusions

By Palmira Pečiuliauskienė

1. Talent, technology, and tolerance are attributes of 'the creative age'. Science is the basis of technology and innovation, and it is particularly important in today's creative society. In 'the creative age', fewer young people seem to be interested in science and technical subjects. These young people belong to Generation Z.

The qualitative research on tenth grade students of Lithuania representing Generation Z shows that more than a half of informants express their interest in Physics. The learners of Generation Z, who like Physics, prefer conducting real rather than virtual laboratory work in Physics. They maintain that performing laboratory work in reality creates conditions for 'lifelike' exploration by touching and fiddling the devices and materials (hands-on dimension), as well as permits experiencing the sense of reality and applying theory in practice.

The learners, who do not like Physics, prefer performing virtual laboratory work. They highlight the uncomplicated and clear technique, fast accomplishment as well as individual speed of conducting virtual laboratory work. Students of Generation Z, who do not take particular interest in Physics, play computer games less frequently than the ones interested in the subject.

2. According to SDT, intrinsic motivation for learning occurs when basic psychological needs (for autonomy, competency, and social relatedness) are satisfied. Digital labs in Physics provide highest satisfaction of new generation students' need for competency, and the lowest satisfaction for the need of interpersonal interactions. Moreover, digital labs in Physics provide better satisfaction with the need for autonomy as compared to the need for interpersonal interactions.

The statistical correlations between Inquiry-based learning (IBL) components and the abilities of information exchange depend on the form of experimental activity (real or virtual). Combining the real and virtual forms in the educational practice of Physics, different IBL components can be stipulated: motivation, feeling of competency, and the feeling of activity autonomy.

The correlation analysis of the main components of the SDT reveals that the strongest and statistically significant correlation is manifested between students' intrinsic motivation for learning Physics and their feeling of competency, as well as between intrinsic motivation for learning Physics and the feeling of autonomy while accomplishing a lab. The correlation between students' intrinsic motivation for learning Physics and their feeling of competency is stronger than the correlation between motivation to learn Physics and the feeling of autonomy while accomplishing a digital lab.

While accomplishing real Physics labs, the basic psychological needs (autonomy, feeling of competency, and interpersonal communication) that are essential for students' intrinsic motivation for learning are also satisfied differently. Real Physics labs tend to mostly satisfy the feelings of autonomy and competency of the new generation students, yet they demand a lot of effort for their accomplishment. The correlation analysis of the main components of SDT demonstrates that while accomplishing real Physics labs, the most prominent and statistically significant correlation was detected between the students' intrinsic motivation for learning Physics and its value/applicability; as well as between the intrinsic motivation for learning and the feeling of competency. The new generation students consider real Physics labs as valuable, since they create preconditions for the students to feel competent.

3. Students learn more positively when teachers support their autonomy rather than control and pressurise them towards a specific way of thinking, feeling, or behaving. Inquiry allows employing an autonomy-supportive style in the classroom, which promotes student engagement as it supports an internal PLOC, an experience of volition, and a sense of choice in students. A well-constructed Physics lab can ensure high level of autonomy in the classroom.

The perspectives of inquiry levels (confirmation inquiry, structured inquiry, guided inquiry, and open enquiry) enable us to deeper understand the peculiarities of motivation for learning Physics. The three characteristics represent the inquiry levels in Physics labs where students could act either independently or not: *problem/question*; *procedures/design*; and *conclusions*. At the level of structured inquiry, students are provided with the problem and procedures of Physics labs. The conclusions are not known before Physics labs. At the level of guided inquiry, students are also provided with the problem and procedures of Physics labs, yet the methods of analysis, communication, and conclusions are for the students to design themselves.

The means of the data of the subscales Interest/Enjoyment, Feeling of Competency, and Effort are higher at the level of structured inquiry Physics labs. However, a statistically significant difference was detected only in one case, namely on the subscale of Feeling of Competency. At the level of structured inquiry, students act less independently. It requires less competency and they feel more competent in less complex situations.

In Physics labs organised at the guided inquiry level, the means of the data of the subscales *Perceived Choice (autonomy)*, and *Value/Usefulness* are higher as compared to the structured inquiry Physics labs. However, this difference is not confirmed statistically. In the guided inquiry labs, students act more independently while selecting the methods of analysis, communication and formulating conclusions. It causes a sense of autonomy and usefulness of Physics labs.

4. Socio-Significant issues (SSI) in science education at school bring science towards socio-cultural environment and reveals goodness and downsides of research and innovation. Intrinsic uncertainty about the future consequences of research and innovation requires a dialogue between students and scientists. SSI activity expands the dialogue from subject-specific communication to interpersonal communication in formal and informal education. SSI activity realised with the help of the 5E model at school enables interpersonal communication of different levels: I- It, I-You, and I-Thou. Students' decision making at SSI activity was based on scientific dilemma and task-related values: attainment value – importance of research and innovation for learner, intrinsic value – enjoyment for learning science, costs of engaging in the task – how much the implementation of research ideas in reality costs. The solving of dilemma problem based on the task-related values required trust and closeness of the project team members including students, teachers, and scientists. The magnitude of the Spearman correlation coefficient indicated the strongest relationship between the variable *This activity was fun to do* (Engagement) and the variable *It is likely that this person and I could become friends if we interacted a lot* (I-You) compared to other associations between the variables of the factor Engagement and variables of the factors I-You and I-Thou.

5. Students employ Mathematics to accomplish Physics labs. Mathematics is deeply woven into both the teaching and practice of Physics. Mathematics is treated as the language of Physics. Motivation for learning Physics should be considered taking into account the peculiarities of motivation for learning Mathematics.

Autonomy support enables the students to get engaged in learning Mathematics. The analysis of autonomy-supportive behaviour and intrinsic motivation has been completed on the basis of *trans-contextual model*, which has a considerable potential to guide research on the transfer of motivation across the classroom work and homework contexts. The correlation analysis reveals that the teacher's autonomy-supportive behaviour based on the interpersonal communication with the students in the classroom is more related to the intrinsic motivation for learning Mathematics. Meanwhile, the teacher's autonomy-supportive behaviour relying on setting Mathematics assignments for students is hardly related to their intrinsic motivation for learning Mathematics.

The correlation analysis reveals that the teacher's individual autonomy-supporting behaviour when working with students in the classroom is more associated with intrinsic motivation for learning mathematics similarly to the homework context.

The benefit students get from learning Mathematics can be social and educational. Lower secondary school students of Lithuania tend to perceive educational benefit from learning Mathematics more than social. The correlation analysis of the variables of extrinsic motivation in the group of educational identification (*Maths is important in learning other subjects*) and the group social identification (*You learn Maths because it is important if you want to get an interesting job*) reveals that educational identification is more associated with intrinsic motivation of school students for learning Mathematics.

The internalised role of Mathematics in learning other subjects (*Maths is important in learning other subjects*) engage students for learning Mathematics more than the internalised role of Mathematics in obtaining entry to a gymnasium or high school (*You learn Maths because it is important in getting entry to a gymnasium or high school*).

Acknowledgement

We would like to thank everyone who has supported and helped us when completing this monograph. Your assistance has been invaluable to us during this process. We also greatly appreciate the financial support of Lithuanian University of Educational Sciences for preparing and publishing of this monograph.

Approval

The monograph MOTIVATION OF NEW GENERATION STUDENTS FOR LEARNING PHYSICS AND MATHEMATICS was approved for publishing by the Council of the Faculty of Education Sciences of the University of Educational Sciences on January 16th 2018 (No. UM7-1).

List of Figures

List of Tables

Bibliography

Abernathy, T. V., & Vineyard, R. N. (2001). Academic competitions in science: What are the rewards for students? *The Clearing House, 74*(5), 269–276.

Agarwal, S., Bansal, S., & Maheshwari, V. K. (2010). *Pragmatism and education*. Retrieved from http://www.scribd.com/doc/30853941/Pragmatism-and-Education.

Alisinanoglu, F., Inan, H. Z., Ozbey, S., & Usak, M. (2012). Early childhood teacher candidates qualifications in science teaching. *Energy Education Science and Technolnology, 4*, 373–390.

Alon, U. (2010). How to build a motivated research group. *Molecular Cell, 37*(2), 151–152.

Anderson, R. D. (2002). Reforming science teaching: What research says about inquiry? *Journal of Science Teacher Education, 13*(1), 1–12.

Arnold, J. C., Kremer, K., & Mayer, J. (2014). Understanding students' experiments—what kind of support do they need in inquiry tasks? *International Journal of Science Education, 36*(16), 2719–2749.

Bajpai, M. (2013). Developing concepts in Physics through virtual lab experiment: An effectiveness study. *Te c h n o L E A R N: An International Journal of Educational Technology, 3*(1), 43–50.

Banchi, H., & Bell, R. (2008). The many levels of inquiry. *Science and Children, 46*(2), 26–29.

Bell, R., Smetana, L., & Binns, I. (2005). Simplifying inquiry instruction. *The Science Teacher, 72*(7), 30–34.

Ben, F. (2010). *Students Uptake of Physics: A Study of Australian and Filipino Physics Students*, School of Education, Faculty of the Proffesions, University of Adelaide.

Bhargava, P., Antonakakis, J., Cunningham, C., & Zehnder, A.T. (2006). Webbased virtual torsion laboratory. *Computer Applications in Engineering Education, 14*(1), 1–8.

Bizzo, N., Kawasaki, C. S., Ferracioli, L., & Leyser da Rosa, V. (2002, July). *Rethinking Science and Technology Education to Meet the Demands of Future Generations in a Changing World*. Proceedings of the International Organization for Science and Technology Education, 10th Symposium. Parana, Brazil: Foz.

Boddy, M., Watson, K., & Aubusson, P. (2003). A trial of the 5E's: A referent model for constructivist teaching and learning. *Research in Science Education, 33*(1), 27–42.

Brophy, J. (2004). Motivating Students to Learn. New Jersey: Lawrence Erlbaum Associates.

Brophy, J. (2008). Developing students' appreciation for what is taught in school. *Educational Psychologist, 43*(3), 132–141.

Brown, P. L., Abell, S. K., Demir, A., & Schmidt, F. J. (2006). College science teachers' views of classroom inquiry. *Science Education, 90*(5), 784–802.

Brownell, S. E., Price, J. V., & Steinman, L. (2013). Science communication to the general public: Why we need to teach undergraduate and graduate students this skill as part of their formal scientific training. *Journal of Undergraduate Neuroscience Education, 12*(1), E6–E10. Retrieved from http://www.ncbi.nlm.nih.gov/pmc/articles/PMC3852879/.

Brozo W. G. (2005). Connecting with students who are disinterested and inexperienced. *Thinking Clasroom, 6*(3), 42–43.

Buber, M. (1970). *I and Thou*. New York: Scribner.

Buck, L. B., Bretz, S. L., & Towns, M. H. (2008).Characterizing the level of inquiry in the undergraduate laboratory. *Journal of College Science Teaching*, 38(1), 52–58.

Bybee, R. W. (2002). Scientific inquiry, student learning, and the science curriculum. In Rodger W. Bybee (Ed.), *Learning Science and the Science of Learning*. USA: National Science Teachers Association.

Bybee, R. W., & McCrae, B. (2011). Scientific literacy and student attitudes: Perspectives from PISA 2006 science. *International Journal of Science Education, 33*(1), 7–26.

Bybee, R. W., Taylor, J. A., Gardner, A., Van Scotter, P., Carlson Powell, J., Westbrook, A., & Landes, N. (2006). *BSCS 5E instructional model: Origins and effectiveness*. A report prepared for the Office of Science Education, National Institutes of Health. Colorado Springs, CO: BSCS.

Byman, R., & Kansanen, P.(2008). Pedagogical thinking in student's mind: A conceptual clarification on the basis of self -determination and volition theories. *Scandinavian Journal of Educational Research, 52*, 603–621.

Byman, R., Lavonen, J., Juuti, K., & Meisalo, V. (2012). Motivational orientations in physics learning: A self-determination theory approach. *Journal of Baltic Science Education, 11*(4), 379–392.

Changeiywo, J. M., Wambugu, P. W., & Wachanga, S. W. (2011). Investigations of students' motivation towards learning secondary school physics through mastery learning approach. *International Journal of Science and Mathematics Education*, *9*(6), 1333–1350.

Chatzisarantis, N. L. D., Hagger, M. S., & Smith, B. (2007). Influences of perceived autonomy support on physical activity within the theory of planned behaviour. *European Journal of Social Psychology*, *37*, 934–954 http://dx.doi.org/10.1002/ejsp.407.

Chiappetta, E. L. (1997). Inquiry-based science: Strategies and techniques for encouraging inquiry in the science classroom. *The Science Teacher*, *64*, 22–26.

Chiappetta, E. L., & Adams, A. D. (2004). Inquiry-based instruction: Understanding how content and process go hand-in-hand with school science. *The Science Teacher*, *71*, 46–51.

Chinn, C. A., & Malhotra, B. A. (2002). Epistemologically authentic inquiry in schools: A theoretical framework for evaluating inquiry tasks. *Science Education*, *86*(2), 175–218.

Cohen, L., Manion, L., & Morrison K. (2007). *Research Methods in Education* (5th edition). London: Routledge Falmer.

Colburn, A. (2000). An inquiry primer. *Science Scope*, *23*(6), 42–44.

Corter, J. E., Nickerson, J. V., Esche, S. K., Chassapis, C., Im, S., & Ma, J. (2007). Constructing reality: A study of remote, hands-on and simulated laboratories, *ACM Transactions on Computer–Human Interaction*, *14*(2), Article 7, 1–27.

Creswell, J. W. (2009). *Research design: Qualitative, quantitative, and mixed methods approaches* (3rd edition). Thousand Oaks, CA: Sage Publications, Inc.

Cross-Bystrom A. (2010). What you need to know about Generation Z. *Imedia Connection*. Retrieved from http://www.imediaconnection.com/content/27425.asp.

Dalgarno, B., & Lee, M. W. (2010). What are the learning affordances of 3D virtual environments? *British Journal of Educational Technology*, *41*(1), 10–32.

Darrah, M., Humbert, R., Finstein, J., Simon, M., & Hopkins, J. (2014). Are virtual labs as effective as hands-on labs for undergraduate Physics? A comparative study at two major universities. *Journal of Science Education and Technology*, *23*(6), 803–814. doi:10.1007/s10956-014-9513-9.

Data of the National Research. (2012). *The material of the research on the achievements of students from Lithuania (2003–2011)*. Retrieved from: http://nec.lt/436/.

DeCharms, R. (1968). *Personal Causation: The Internal Affective Determinants of Behaviour*. New York: Academic.

Deci, E. L. (1975). *Intrinsic Motivation*. New York: Plenum.

Deci, E. L., & Ryan, R. M. (1985). *Intrinsic Motivation and Self-determination in Human Behaviour*. New York: Plenum.

Deci, E. L., & Ryan, R. M. (1991). A motivational approach to self: Integration in personality. In R. Dienstbier (Ed.), *Perspectives on Motivation. Nebraska Symposium on Motivation, 38* (pp. 237–288). Lincoln, NE: University of Nebraska Press.

Deci, E. L., & Ryan, R. M. (2000a). The "what" and "why" of goal pursuits: Human needs and the self-determination of behaviour. *Psychological Inquiry, 11*(4), 227–268.

Deci, E. L., & Ryan, R. M. (2000b). Intrinsic and extrinsic motivations: Classic definitions and new directions. *Contemporary Educational Psychology, 25*, 54–67.

Deci, E. L., & Ryan, R. M. (2002). *Handbook of Self-determination Research*. Rochester, NY: University of Rochester Press.

Deci, E. L., & Ryan, R. M. (2008). Self-determination theory: A macro theory of human motivation, development and health. *Canadian Psychology, 49*(3), 182–185.

Deci, E. L, & Ryan, R. M. (2012a). *Motivation, personality, and development within embedded social contexts: An overview of self-determination theory*. In E. Deci, & R. Ryan (Eds.), *The Oxford Handbook of Human Motivation* (pp. 85–107). New York: Oxford University Press.

Deci, E. L., & Ryan, R. M. (2012b). Self-determination theory. In P. A. M. Van Lange, A. W. Kruglanski, & E. T. Higgins (Eds.), *The Handbook of Theories of Social Psychology*, 1 (pp. 416–437). Retrieved from http://books.google.lt/books?id=1s-tMyLamcC&pg=PA416&hl=lt&source=gbs_toc_r&cad=4#v=onepage&q&f=false.

Deci, E. L., Eghrari, H., Patrick, B. C., & Leone, D. (1994). Facilitating internalization: The self-determination theory perspective. *Journal of Personality, 62*, 119–142.

Deci, E. L., Ryan, R. M., & Guay, F. (2013). Self-determination theory and actualization of human potential. In D. McInerney, H. Marsh, R. Craven, & F. Guay (Eds.), *Theory Driving Research: New Wave Perspectives on Self Processes and Human Development* (pp. 109–133). Charlotte, NC: Information Age Press.

Deci, E., & Ryan, R. (2002). *Overview of self-determination theory: An organismic dialectical perspective*. In E. Deci, & R. Ryan (Eds.), *Handbook of Self-Determination Research*.

Deci, E. L., Ryan, R. M., & Williams, G. C. (1996). Need satisfaction and the self-regulation of learning. *Learning and Individual Differences*, *8*, 165–183.

Deci, E. L., Vallerand, R. J., Pelletier, L. G., & Ryan, R. M. (1991). Motivation and education: The self-determination perspective. *The Educational Psychologist*, *26*, 325–346.

Dierking, L. D., Falk, J. H. (2016). 2020 vision: Envisioning a new generation of stem learning research. *Cultural Studies of Science Education*, *11*(1), 1–10.

Downes, S. (2007). "What Connectivism Is". Retrieved from http://www.downes.ca/post/38653.

Eccles, J. S., & Wigfield, A. (2002). Motivational beliefs, values, and goals. *Annual Review of Psychology*, *53*(1), 109–132.

Edelson, D. C. (2001). Learning-for-use: A framework for the design of technology-supported inquiry activities. *Journal of Research in Science Teaching*, *38*(3), 355–385.

Edelson, D. C., Gordin, D. N., & Pea, R. D. (1999). Addressing the challenges of inquiry-based learning through technology and curriculum design. *Journal of the learning sciences*, *8*(3–4), 391–450.

Education Management Information System (EMIS). (2016). *Statistics of Basic Education of 2016–2017*. Ministry of Education and Science of the Republic of Lithuania. Retrieved from http://svis.emokykla.lt/.

Ehrlinger, J., Gilovich, T., & Ross, L. (2005). Peering into the bias blind pot: People's assessments of bias in themselves and others. *Personality and Social Psychology Bulletin*, *31*(5), 680–692.

Eow, Y. L., Wan Zah, B. W. A., Roselan, B., & Rosnaini, M. (2010). Stability of the Intrinsic Motivation Inventory (IMI) for the use of Malaysian form one students in ICT literacy class. *Eurasia Journal of Mathematics, Science & Technology Education*, *6*(3), 215–226.

Espeja, A. G., & Lagarón, D. C. (2015). Socio-scientific issues (SSI) in initial training of primary school teachers: Pre-service teachers' conceptualization of SSI and appreciation of the value of teaching SSI. *Procedia – Social and Behavioural Sciences*, *196*, 80–88.

Etkina, E., Karelina, A., Ruibal-Villasenor, M., Rosengrant, D., Jordan, R., & Hmelo-silver, C. E. (2010). Design and reflection help students develop scientific abilities: Learning in introductory physics laboratories. *The Journal of the Learning Sciences*, *19*(1), 54–98.

Evagorou, M., Jimenez-aleixandre, M. P., & Osborne, J. (2012). "Should we kill the grey squirrels?" A study exploring students' justifications and decision-making. *International Journal of Science Education*, *34*(3), 401–428.

Exner, N. (2014). Research information literacy: Addressing original researchers' needs. *Journal of Academic Librarianship*, *40*(5), 460–466. doi:10.1016/j.acalib.2014.06.006.

Field, A. (2000). *Discovering Statistics Using SPSS for Windows*. London – Thousand Oaks – New Delhi: Sage Publications.

Fiore, S. M. (2008). Interdisciplinarity as team work: How the science of teams can inform team science. *Small Group Res*earch, *39*(3), 251–277.

Fisher, H. R., & Horstendal, M. (1997). Motivation and learning physics. *Research and Science Education*, *27*(3), 411–424.

Flick, L. B. (1995). Complex instruction in complex classrooms: A synthesis of research in inquiry teaching methods and explicit teaching strategies. Paper presented at the annual Meeting of the National Association for Research in Science Teaching. San Francisco, CA.

Florida, R. L. (2002). *The Rise of the Creative Class: And How it's Transforming Work, Leisure, Community and Everyday Life*. New York: Basic Books.

Fortus, D. (2014). Attending to affect. *Journal of Research in Science Teaching*, *51*(7), 821–835.

Freedman, M. P. (1997). Relationship among laboratory instruction, attitude toward science, and achievement in science knowledge. *Journal of Research in Science Teaching*, *34*(4), 343–357. doi:10.1002/(SICI)1098-2736(199704)34:4<343::AID-TEA5>3.0.CO;2-R.

Freedman, K. (2004). Editorial: Becoming a researcher in art education: Forming research questions. *Studies in Art Education*, *45*(2), 99–100.

Furberg, A. (2010). *Scientific Inquiry in Web-based Learning Environments. Exploring Technological, Epistemic and Institutional Aspects of Students' Meaning Making*. Oslo: Oslo University Press.

Furtak, E.M., Kunter, M. (2012). Effects of autonomy-supportive teaching on student learning and motivation. *Journal of Experimental Education*, *80*, 284–316.

Gagné, M., & Deci, E. L., (2005). Self-determination theory and work motivation. *Journal of Organizational Behaviour*, *26*(4), 331–362.

Gardner, P. L. (1975). Attitudes to science: A review. *Studies in Science Education*, *1*(1), 1–41.

Good, R. G., Wandersee, J. H., & St. Julien, J. (1993). Cautionary notes on the appeal of the new ism (*constructivism*) in science education. In K. Tobin

(Ed.), *The Practice of Constructivism in Science Education* (pp. 71–87). Hillsdale, NJ: Erlbaum.

Goodman, P., Rose, J., & Furcon, J. (1970). A Comparison of motivational antecedents of the work performance of scientists and engineers. *Journal of Applied Psychology, 54*(6), 491–495.

Gormally, C., Brickman, P., Hallar, B., & Armstrong, N. (2009). Effects of inquiry-based learning on students' science literacy skills and confidence. *International Journal for the Scholarship of Teaching and Learning, 3*(2), 16.

Gravetter, F., & Wallnau, L. (2014). *Essentials of Statistics for the Behavioral Sciences* (8th ed.). Belmont, CA: Wadsworth.

Greeno, J., Collins, A., & Resnick, L. B. (1996). Cognition and learning. In R. Calfee & D. Berliner (Eds.), *Handbook of Educational Psychology*. New York: Macmillan.

Guay, F., Lessard, V., & Dubois, P. (2016). How can we create better learning contexts for children? Promoting students' autonomous motivation as a way to foster enhanced educational outcomes. In J. Wang, W.C., Liu, & R.M. Ryan (Eds.), *Building Autonomous Learners* (pp. 83–106). Singapore: Springer.

Hagger, M. S., & Chatzisarantis, N. L. D. (2012). Transferring motivation from educational to extramural contexts: A review of the trans-contextual model. *European Journal of Psychology of Education, 27*, 195–212.

Hagger, M., Sultan, S., Hardcastle, S., & Chatzisarantis, N. (2015). Perceived autonomy support and autonomous motivation toward mathematics activities in educational and out-of-school contexts is related to mathematics homework behaviour and attainment. *Contemporary Educational Psychology, 41*, 111–123.

Harwood, W. S. (2004a). A new model for inquiry. *Journal of College Science Teaching, 33*(7), 29–33.

Harwood, W. S. (2004b). An activity model for scientific inquiry. *The Science Teacher*, 71(1), 44–46.

Healey, M. (2005). Linking research and teaching: Exploring disciplinary spaces and the role of inquiry-based learning. In R. Barnett (Ed.), *Reshaping the University: New Relationships Between Research, Scholarship and Teaching* (pp. 67–78). London: McGraw Hill and Open University Press.

Healey, M. & Roberts, J. (Eds.). (2004). *Engaging Students in Active Learning: Case Studies in Geography, Environment and Related Disciplines*. Cheltenham: Geography Discipline Network and School of Environment, University of Gloucestershire.

Heider, E. (1958). *The Psychology of Interpersonal Relations*. New York: Wiley.

Heinich, R., Molenda, M., Russell, J. D., & Smaldino, S. E. (2002). *Instructional Media and Technologies for Learning*. Upper Saddle River, NJ: Pearson Education.

Herron, M. D. (1971).The nature of scientific inquiry. *School Review, 79*(2), 171–212.

Hidi, S. (2006). Interest: A motivational variable with a difference. *Educational Research Review, 7*, 323–350.

Hidi, S. (2011). Linking neuroscientific research to interest research: Theoretical and empirical considerations. In P. A. O'Keefe & I. Plante (Chairs), Developments in interest theory and research. Symposium conducted at the annual meeting of the American Educational Research Association, New Orleans, LA.

Hidi, S., & Renninger, K. A. (2006). The four-phase model of interest development. *Educational Psychologist, 41*, 111–127.

Higgins, S. E. (2010). The impact of interactive whiteboards on classroom interaction and learning in primary schools in the UK. In S. Bax et al. *Interactive Whiteboards for Education: Theory, Research and Practice* (pp. 86–101). Hershey, PA: IGI Global.

Higgins, S. E., Wall, K., & Smith, H. (2005). The visual helps me understand the complicated things: Pupil views of teaching and learning with interactive whiteboards. *British Journal of Educational Technology, 36*(5), 851–867.

Hofstein, A., & Lunetta, V. (2004). The laboratory in science education: Foundations for the twenty-first century. *Science Education, 88*(1), 28–54. doi:10.1002/sce.10106.

Hughes, P. W., & Ellefson, M. R. (2013). Inquiry-based training improves teaching effectiveness of biology teaching assistants. *PLoS ONE, 8*(10), e78540. http://doi.org/10.1371/journal.pone.0078540.

Hurtz, G. M., & Williams, K. J. (2009). Attitudinal and motivational antecedents of participation involuntary employee development activities. *Journal of Applied Psychology, 94*(3), 635–653.

Igwebuike, T., & Oriaifo, S. (2012). Effect of a constructivist instructional strategy on affective outcomes by integrated science students. *International Review of Contemporary Learning Research, 3*(1), 1–10.

Ince, E., Kirbaslar, F. G., Yolcu, E., et al. (2014). 3-Dimensonal and interactive Istanbul university virtual laboratory based on active learning methods. *Tojet: The Turkish Online Journal of Educational Technology, 13*(1), 1–20.

Jaakkola, T., & Nurmi, S. (2008). Fostering elementary school students' understanding of simple electricity by combining simulation and laboratory activities. *Journal of Computer Assisted Learning, 24*, 271–283. doi:10.1111/j.1365-2729.2007.00259.x.

Jang, H., Kim, E. J., Reeve, J. (2012). Longitudinal test of self-determination theory's motivation mediation model in a naturally occurring classroom context. *Journal of Educational Psychology, 104*, 1175–1188. http://dx.doi.org/10.1037/a0028089.

Jang, H., Reeve, J., & Halusic, M. (2016). A new autonomy-supportive way of teaching that increases conceptual learning: Teaching in students' preferred ways. *The Journal of Experimental Education, 84*(4), 686–701. doi:http://dx.doi.org/10.1080/00220973.2015.1083522.

Jobrack, B. (2011). *Tyranny of the Textbook: An Insider Exposes How Educational Materials Undermine Reforms*. Lanham, MD: Rowman & Littlefield.

Jurik, V., Gröschner, A., & Seidel, T. (2014). Predicting students' cognitive learning activity and intrinsic learning motivation: How powerful are teacher statements, student profiles, and gender? *Learning and Individual Differences, 32*, 132–139.

Juuti, K., Lavonen, J., Aksela, M., & Meisalo, V. (2009). Adoption of ICT in science education: a case study of communication channels in a teachers' professional development project. *Eurasia Journal of Mathematics, Science and Technology Education, 5*, 103–118.

Kim, M., Tan, A. L., & Talaue, F. T. (2013). New vision and challenges in inquiry-based curriculum change in Singapore. *International Journal of Science Education, 35*(2), 289–311.

Klahr, D., Triona, L. M., & Williams, C. (2007). Hands on what? The relative effectiveness of physical versus virtual materials in an engineering design project by middle school children. *Journal of Research in Science Teaching, 44*(1), 183–203.

Krapp, A., & Prenzel, M. (2011). Research on interest in science: Theories, methods, and findings. *International Journal of Science Education, 33*(1), 27–50.

Lang, J. (2012). Comparative study of hands-on and remote physics labs for first year university level Physics students. *Transformative Dialogues: Teaching & Learning Journal, 6*(1), 1–25.

Lavonen, J., Gedrovics, J., Byman, R., Meisalo, V., Juuti, K., & Uitto, A. (2008). Students' motivational orientations and career choice in science and technology: A survey in Finland and Latvia. *Journal of Baltic Science Education, 7*(2), 86–103.

Lawlor, J., Marshall, K., & Tangney, B. (2015). Bridge21 – exploring the potential to foster intrinsic student motivation through a team-based, technology-mediated learning model. *Technology, Pedagogy and Education, 25*(2), 187–206.

Lemke, J. L. (2001). Articulating communities: Sociocultural perspectives on science education. *Journal of Research in Science Teaching, 38*, 296–316.

Levri, E. P., & Levri, M. A. (2003). Hot salsa: A laboratory exercise exploring the scientific method. *The American Biology Teacher, 65*(5), 372–377.

Linn, M. C., Davis, E. A., & Bell, P. (2004). Inquiry and technology. In M. C. Linn, E. A. Davis, & P. Bell (Eds.), *Internet Environments for Science Education* (pp. 3–28). Mahwah, NJ: Lawrence Erlbaum Associates.

Llewellyn, D. J. (2011). *Differentiated Science Inquiry*. Thousand Oaks, CA: A Sage Company Ltd.

Lnag, L. (2012). Comparative study of hands-on and remote physics labs for first year university level physics students. *Transformative Dialogues: Teaching & Learning Journal, 6*(1), 1–25.

Long, C. Y., & Jiar, Y. K. (2014). Mathematical thinking and physics achievement of secondary school students. *Sains Humanika, 2*(4), 231–237.

Lotrecchiano, G. R., Mallinson, T. R., Leblanc-Beaudoin, T., Schwartz, L. S., Lazar, D., & Falk-Krzesinski, H. J. (2016). Individual motivation and threat indicators of collaboration readiness in scientific knowledge producing teams: A scoping review and domain analysis. *Heliyon, 2*(5), e00105. htttp://doi.org/10.1016/j.heliyon.2016.e00105.

Loukomies, A. (2013). Enhancing Students' Motivation towards School Science with an Inquiry-Based Site Visit Teaching Sequence: A Design-Based Research Approach (Doctoral dissertation, University of Helsinki, 2013).

Loukomies, A. Pnevmatikos, D., Lavonen, J, Spyrtou, A., Byman, R, Kariotoglou, P., & Juuti, K. (2013). Promoting students' interest and motivation towards science learning: the role of personal needs and motivation orientations. *Research in Science Education, 43*(6), 2517–2539.

Lowe, D., Newcombe, P, & Stumpers, B. (2013). Evaluation of the use of remote laboratories for secondary school science education. *Research in Science Education, 43*, 1197–1219.

Maglaughlin, K. L., & Sonnenwald, D. H. (2005). Factors that impact interdisciplinary natural science research collaboration in academia. In *Proceedings of ISSI 2005: 10th International Conference of the International Society for Scientometrics and Informetrics*. vol. 2, 499–508.

Margolis, H., & McCabe, P. (2006). Improving self-efficacy and motivation: What to do, what to say. *Intervention in School and Clinic, 41*(4), 218–227.

Mattar, J. A. (2010). *Constructivism and connectivism in education technology: Active, situated, authentic, experiential, and anchored learning.* Retrieved from https://pdfs.semanticscholar.org/d2c3/a33d7c7fcc5822ff0449 4863490045f8758e.pdf.

Mayer, R. E. (2004). Should there be a three-strikes rule against pure discovery learning? The case for guided methods of instruction. *American Psychologist, 59*(1), 14–19. doi:10.1037/0003-066X.59.1.14.

McAuley, E., Duncan, T., & Tammen, V. V. (1987). Psychometric properties of the Intrinsic Motivation Inventory in a competitive sport setting: A confirmatory factor analysis. *Research Quarterly Exercise and Sport, 60*(1), 48–58.

McCrindle, M., & Wolfinger, E. (2010). *The ABC of XYZ: Understanding the Global Generation*. Kensington: University of New South Wales Press Ltd.

Meltzer, D. E. (2002). The relationship between mathematics preparation and conceptual learning gains in physics: a possible 'hidden variable' in diagnostic pre test scores. *The American Journal of Physics, 70*(12), 1259–1268.

Mercer-Mapstone, L. & Kuchel, L. (2015). Teaching scientists to communicate: Evidence-based assessment for undergraduate science education, *International Journal of Science Education, 37*(10), 1613–1638. doi :10.1080/09500693.2015.1045959.

Minner, D. D., Levy, A. J., & Century, J. (2010). Inquiry-based science instruction – what is it and does it matter? Results from a research synthesis years 1984–2002. *Journal of Research in Science Teaching, 47*(4), 474–496. doi:10.1002/tea.20347.

National Academy of Science (NAS) (2015). *Enhancing the Effectiveness of Team Science*. Washington, DC: The National Academies of Science, Committee on the Science of Team Science; Board on Behavioural, Cognitive, and Sensory Sciences; Division of Behavioural and Social Sciences and Education; National Research Council.

National Research Council (NRC). (1996). *National Science Education Standards*. Washington, DC: National Academy Press.

National Research Council (NRC). (2000). *Inquiry and the National Science Education Standards: A Guide for Teaching and Learning*. Washington, DC: National Academy Press.

National Research Council (NRC). (2012). *A Framework for K-12 Science Education: Practices, Crosscutting Concepts, and Core Ideas*. Washington, DC: The National Academies Press.

Melnikė, E., Jaruševičiūtė, V., Skripkienė, R., Baliutavičiūtė, G., Bilinskas, B., & Valavičienė, N. (2004). National student achievement test. [In

Lithuanian]. Retrieved from file:///C:/Users/pp/Downloads/2004_Mokin_l_bendrieji10%20(1).pdf.

Melnikė, E., Jaruševičiūtė, V., Skripkienė, R., Baliutavičiūtė, G., Bilinskas, B., and Valavičienė, N. (2004). National student achievement test. [In Lithuanian]. Retrieved from https://www.upc.smm.lt/ekspertavimas/tyrimai/bazes/failai/2008_Mokin_droji_dalis.pdf.

Neo, M., & Neo, T. K. (2009). Engaging students in multimedia-mediated constructivist learning-students' perceptions. *Educational Technology and Society*, *12*(2), 254–266.

Nedic, Z., Machotka, J., & Nafalski, A. (2003). Remote laboratories versus virtual and real laboratories. *Proceedings of the 33rd ASEE/IEEE Frontiers in Education Conference*, vol. 1, T3E-1–T3E-6). San Diego, CA: ASEE/IEEE. Retrieved from http://www.icee.usm.edu/ICEE/conferences/FIEC2003/papers/1077.pdf.

Ng B. L., Liu W. C., & Wang J. C. (2015). Student motivation and learning in mathematics and science: a cluster analysis. *International Journal of Science and Mathematics Education*, *14*, 1359–1376. doi:10.1007/s10763-015-9654-1.

Nolen, S. (2003). Learning environment, motivation, and achievement in high school science. *Journal of Research in Science Teaching*, *40*(4), 347–368.

Organisation for Economic Co-operation and Development (OECD).(2007). *PISA 2006: Science Competencies for Tomorrow's World: Volume 1: Analysis*. Paris: OECD.

Ohle, A., Boone, W. J., & Fischer, H. S. (2015). Investigating the impact of teachers' Physics Ck on students outcomes. *International Journal of Science and Mathematics Education*, *13*, 1211–1233.

Oidov, L., Tortogtokh, U., & Purevdagva, E. (2012). Virtual laboratory for physics teaching. *International Proceedings of Economics Development & Research*, *37*, 319–323.

Okada, A. (2013). Scientific literacy in the digital age: Tools, environments and resources for co-inquiry. *European Scientific Journal*, *4*, 263–274.

Okada, A. (Ed.). (2016). *Engaging Science: Innovative Teaching for Responsible Citizenship*. Milton Keynes: The Open University – Knowledge Media Institute.

Okada, A., Young, G., & Sanders, J. (2015). Fostering communities of practices for teachers' professional development integrating OER and MOOC, EC-TEL. The 10th European Conference on Technology Enhanced Learning. September 15nd–18rd, 2015. Spain, Toledo.

Olusola, O. O., & Rotimi, C. O. (2012). Attitudes of students towards the study of physics in College of Education Ikere Ekiti, Ekiti State, Nigeria. *American International Journal of Contemporary Research*, *2*(12), 86–89.

Ormerod, M. B., & Duckworth, D. (1975). *Pupils' Attitudes to Science*. Slough: NFER.

Osborne, J., & Dillon, J. (2008). *Science Education in Europe: Critical Reflections*. London: The Nuffield Foundation.

Osborne, J., Simon, Sh., & Collins, S. (2003) Attitudes towards science: A review of the literature and its implications. *International Journal of Science Education*, *25*(9), 1049–1079. doi:10.1080/0950069032000032199 .Owen, R., Macnaghten, P., & Stilgoe, J. (2012). Responsible research and innovation: From science in society to science for society, with society. *Science & Public Policy (SPP)*, *39*(6), 751–760. doi:10.1093/scipol/scs093.

Palmer, D. (2005). A motivational view of constructivist informed teaching. *International Journal of Science Education*, *27*(15,16), 1853–1881.

Pardee, R. L. (1990). *Motivation theories of Maslow, Herzberg, McGregor & McClelland. A literature review of selected theories dealing with job satisfaction and motivation*. Retrieved from http://files.eric.ed.gov/fulltext/ED316767.pdf.

Patton, M. Q. (2002). *Qualitative Research & Evaluation Methods*. Thousand Oaks, CA: Sage Publications.

Peciuliauskiene, P. (2014). E-learning and motivation for learning physics at school: the case of generations Y and Z. *DIVAI 2014: 10th International Scientific Conference on Distance Learning in Applied Informatics: Conference Proceedings*: Štúrovo, Slovakia, May 5–7, 2014. Štúrovo, 2014. ISBN 9788074784972, pp. 441–451.

Peciuliauskiene, P. (2015a). Interpersonal communication of school students in physical experimental activity: the aspect of real and digital labs. *Academic Journal of Interdisciplinary Studies*, *4*(3), 679–688.

Peciuliauskiene, P. (2015b). Promotion of intrinsic motivation of new generation learners for learning Physics by digital Physics labs. *Society, Integration, Education*: Proceedings of the International Scientific Conference, May 22nd–23rd, 2015. Rezekne Higher Education Institution. Faculty of Education and Design. Velta Lubkina, & Svetlana Usca (Eds.). Rezekne: Rêzeknes Augstskola.

Petersen, S., & Wulff, P. (2017). The German Physics Olympiad—identifying and inspiring talents. *European Journal of Physics*, *38*(3), 25–30.

Petersson, E., Andersson, A., & Säljö, R. (2013). Exploring nature through virtual experimentation. *Nordic Journal of Digital Literacy*, *3*, 139–156.

Piaget, J. (1978). *The Development of Thought: Equilibration of Cognitive Structures*. Oxford: Blackwell.

Pintrich, P. R. (2003). A motivational science perspective on the role of student motivation in learning and teaching contexts. *Journal of Educational Psychology, 95*(4), 667–686.

Pintrich, P. R., Roeser, R. W., & de Groot, E. A. M. (1994). Classroom and individual differences in early adolescents' motivation and self-regulated learning. *The Journal of Early Adolescence, 14*, 139–161.

Potvin, P., & Hasni, A. (2014). Interest, motivation and attitude towards science and technology at K-12 levels: A systematic review of 12 years of educational research. *Studies in Science Education, 50*(1), 1–45.

Pyle, E. J. (1995). *Motivation, action, and social cognitive development in informal science education venues: Convergence with the classroom* (Doctoral dissertation, University of Georgia: Athens).

Melnikė, E., Jaruševičiūtė, V., Skripkienė, R., Baliutavičiūtė, G., Bilinskas, B., & Valavičienė, N. (2012). *Questionnaire about mathematics for 8th grade students.* [In Lithuanian]. Retrieved from http://nec.lt/failai/4269_2012NMPT_8klase_Klausimynas_matematika.pdf.

Rayan, R., & Deci, E. L. (2000). Intrinsic and extrinsic motivations: Classic definitions and new directions. *Contemporary Educational Psychology, 25*, 54–67.

Redish, E. (2004). A theoretical framework for physics education research: Modeling student thinking. In E. F. Redish and M. Vicentini (Eds.), *Proceedings of the International School of Physics, "Enrico Fermi" Course CLVI*, Varenna, Italy, August 2003, E. F. Redish & M. Vicentini (Eds.) (IOS Press, Amsterdam, 2004, 1–65).

Redish, E. F., & Gupta, A. (2010). Making Meaning with Maths in Physics: A Semantic Analysis. In D. Raine, C. Hurkett, & L. Rogers (Eds.), Physics Community and Cooperation: GIREP-EPEC & PHEC. 2009 International Conference, Leicester, UK, August 20, 2009, 244–260 (2010).

Redish, E. F., & Kuo, E. (2015). Language of physics, language of math: Disciplinary culture and dynamic epistemology. *Science & Education, 24*(5–6), 561–590. doi:10.1007/s11191-015-9749-7.

Reeve, J. (2002). Self-determination theory applied to educational settings. In E. Deci & R. M. Ryan (Eds.), *Handbook of Self-determination Research* (pp. 183–203). Rochester, NY: University of Rochester Press.

Reeve, J., Bolt, E., & Cai, Y. (1999). Autonomy-supportive teachers: How they teach and motivate students. *Journal of Educational Psychology, 91* (1999), 537–548. http://dx.doi.org/10.1037/0022-0663.91.3.537.

Reeve, J., & Jang, H. (2006). What teachers say and do to support students' autonomy during a learning activity. *Journal of Educational Psychology*, *98*, 209–218.

Reeve, J., & Jang, H. (2009). Why teachers adopt a controlling motivating style toward students and how they can become more autonomy supportive. *Educational Psychologist*, *44*(3), 159–175.

Reeve, J., & Lee, W. (2014). Students' classroom engagement produces longitudinal changes in classroom motivation. *Journal of Educational Psychology*, *106*, 527–540. doi:10.1037/a0034934.

Reeve, J., Nix, G., & Hamm, D. (2003). Testing models of the experience of self-determination in intrinsic motivation and the conundrum of choice. *Journal of Educational Psychology*, *95*, 375–392.

Reeve, J., & Tseng, C. (2009a). *Adding Students' Voice to the Conceptualization of Engagement in Learning Activities*. Unpublished manuscript, University of Iowa, Iowa City.

Reeve, J., & Tseng, C. (2009b). *Teachers' Motivating Styles Toward Students: The Psychobiology of Being Controlled*. Unpublished manuscript, University of Iowa, Iowa City.

Reiff, R., Harwood, W. S., & Phillipson, T. (2002). A scientific method based upon research scientists' conceptions of scientific inquiry. Paper presented at the AETS, Charlotte, NC.

Reeve, J., Deci, E. L., & Ryan, R. M. (2004). Self-determination theory: A dialectical framework for understanding the sociocultural influences on student motivation. In D. M. McInerney, & S. Van Etten (Eds.), *Research on Sociocultural Influences on Motivation and Learning: Big Theories Revisited*, *4* (pp. 31–59). Greenwich, CT: Information Age Press.

Reeve, J., Jang, H., Carrell, D., Barch, J., & Jeon, S. (2004). Enhancing high school students' engagement by increasing their teachers' autonomy support. *Motivation and Emotion*, *28*, 147–169.

Rezba, R. I., Auldridge, T. & Rhea, L. (1999). *Teaching & learning the basic science skills*. Retrieved from www.pen.k12.va.us/ VDOEIinstiruc-tion/ TLBSSGuide.doc.

Rietveld, T., & Van Hout, R. (1993). *Statistical Techniques for the Study of Language and Language Behaviour*. Berlin – New York: Mouton de Gruyter.

Riga F., Winterbottom M., Harris E., & Newby L. (2017). Inquiry-based science education. In Taber, K. S., Akpan, B. (Eds.), *Science Education. New Directions in Mathematics and Science Education*. Rotterdam: SensePublishers.

Rissing, S., W., & Cogan, J. G. (2009). Can an inquiry approach improve college student learning in a teaching laboratory? *CBE Life Sciences Education, 8*(1), 55–61.

Rocard, M., Csermely, P., Jorde, D., Lenzen, D., Walberg-Henriksson, H., & Hemmo, V. (2007). *Rocard report: "Science Education Now: A New Pedagogy for the Future of Europe"*. EU 22845, European Commission.

Rosen, L. D. (2012). *iDisorder. Understanding Our Obsession with Technology and Overcoming its Hold on Us*. New York: Palgrave Macmillan.

Roth, G., Assor, A., Kanat-Maymon, Y., & Kaplan, H. (2007). Autonomous motivation for teaching: How self-determined teaching may lead to self-determined learning. *Journal of Educational Psychology, 99*, 761–774.

Rosen, Y.(2009).The effects of an animation-based on-line learning environment on transfer of knowledge and on motivation for science and technology learning. *Journal of Educational Computing Research, 40*(4), 451–467, https://doi.org/10.2190/EC.40.4.d.

Rutherford, F. J. (1964). The role of inquiry in science teaching. *Journal of Research in Science Teaching, 2*, 80–84.

Ryan, R. M. (1982). Control and information in the intrapersonal sphere: An extension of cognitive evaluation theory. *Journal of Personality and Social Psychology, 43*, 450–461.

Ryan, R.M., & Deci, E. L. (2000). Intrinsic and extrinsic motivations: Classic definitions and new directions. *Contemporary Educational Psychology, 25*, 54–67.

Ryan, R. M., & Deci, E. L. (2002). An overview of self-determination theory: An organismic–dialectical perspective. In E. L. Deci & R. M. Ryan (Eds.), *Handbook of Self-determination Research* (pp. 3–33). Rochester: The University of Rochester Press.

Ryan, R. M., & Deci, E. L. (2004). Autonomy is no illusion: Self-determination theory and the empirical study of authenticity, awareness, and will. In J. Greenberg, S. L. Koole, & T. Pyszczynski (Eds.), *Handbook of Experimental Existential Psychology* (pp. 449–479). New York: Guilford Press.

Ryan, R. M., & Deci, E. L. (2009). Promoting self-determined school engagement. In K. R. Wentzel & A. Wigfield (Eds.), *Handbook of Motivation at School*. New York: Taylor & Francis.

Ryan, R. M., Koestner, R., & Deci, E. L. (1991). Varied forms of persistence: When free-choice behaviour is not intrinsically motivated. *Motivation and Emotion, 15*, 185–205.

Ryan, R. M., & Patrick, H. (2001). The classroom social environment and changes in adolescents'motivation and engagement during middle school. *American Educational Research Journal, 38*, 437–460.

Savelsbergh, E. R., Prins, G. T.; Rietbergen, C., Fechner, S., Vaessen, B. E., Draijer, J. M., Bakker, A. (2016). Effects of innovative science and mathematics teaching on student attitudes and achievement: A meta-analytic study. *Educational Research Review, 19*, 158–172.

Schroeder, C., Scott, T., Tolson, H., Huang, T., & Lee, Y. (2007). A meta-analysis of national research: Effects of teaching strategies on student achievement in science in the United States. *Journal of Research in ScienceTeaching, 44*(10), 1436–1460.

Schunk, D. H., Pintrich, P. R., & Meece, J. L. (2007). *Motivation in Education.* Upper Saddle River, NJ: Pearson Education.

Schunk, D. H., Pintrich, P. R., & Meece, J. L. (2008). *Motivation in Education: Theory, Research and Application* (3rd ed.). Upper Saddle River, NJ: Merrill.

Schukajlow, S., Rakoczy, K., & Pekrun, R. (2017). Emotions and motivation in mathematics education: Theoretical considerations and empirical contributions. *ZDM Mathematics Education, 49*(3), 307–322.

Schwab, J.J. (1962). *The Teaching of Science as Inquiry*. In J. J. Schwab & P. E Brandwein (Eds.), *The Teaching of Science* (pp. 3–103). Cambridge, MA: Harvard University Press.

Delhaxhe, A., Forsthuber, B. (Coord.), Motiejunaite, A., de Almeida Coutinho, A. S., Baïdak, N., & Horvath, A. Science *education* in Europe: National *policies, practices and research*. (2011). Retrieved from http://eacea.ec.europa.eu/education/eurydice.

Seemiller, C., & Grace, M. (2016). *Generation Z Goes to College*. New York: Jossey-Bass.

Sherborne, T., et al. (2014). *ENGAGE: Equipping the next generation for active engagement in science*. Retrieved from http://cordis.europa.eu/project/rcn/111469_en.html.

Shumow, L. Schmidt, J. A., & Zaleski, D. J. (2013). Multiple perspectives on student learning, engagement, and motivation in high school biology labs. *The High School Journal, 96*(3), 232–252.

Siemens, G. (2004). *Connectivism: A theory for the digital age*. Retrieved from http://www.elearnspace.org/Articles/connectivism.htm.

Stahl, B. C. (2013). Responsible research and innovation: The role of privacy in an emerging framework. *Science & Public Policy (SPP), 40*(6), 708–716.

Stefanou, C. R., Perencevich, K. C., DiCintio, M., & Turner, J. C. (2004). Supporting autonomy in the classroom: Ways teachers encourage student decision making and ownership. *Educational Psychologist, 39*, 97–110. doi:10.1207/s15326985ep3902 2.

Stokols, D., Hall, K. L., Taylor, B. K., & Moser, R. P. (2008a).The science of team science: Overview of the field and introduction to the supplement. *American Journal of Preventive Medicine 35*(Suppl. 2), S77–S89.

Stokols, D., Misra, S., Moser, R. P., Hall, K. L., & Taylor, B. K. (2008b). The ecology of team science: Understanding contextual influences on transdisciplinary collaboration. *American Journal of Preventive Medicine 35*(2), S96–S115.

Swarat, S., Ortony, A., & Revelle, W. (2012). Activity matters: Understanding student interest in school science. *Journal of Research in Science Teaching, 49*(4), 515–537.

Tam, M. (2000). Constructivism, instructional design, and technology: Implications for transforming distance learning. *Educational Technology & Society, 3*(2), 50–60.

Thornton, R. K. (2008). Effective learning environments for computer supported instruction in the physics classroom and laboratory. In: Vicentini, M & Sassi, E (Eds.) *Connecting Research in Physics Education with Teacher Education.* (Vol. 2) International Commission on Physics Education, pp. 1–21.

Tsigilis, N., & Theodosiou, A. (2003). Temporal stability of the intrinsic motivation inventory. *Perceptual and Motor Skills, 97*(1), 271–280.

Tytler, R., Osborne, J., Williams, G., Tytler, K., & Cripps C. J. (2008). *Opening up Pathways: Engagement in STEM across the Primary–Secondary School Transition.* Canberra: Australian Department of Education, Employment and Workplace Relations. Retrieved from https://docs.education.gov.au/system/files/doc/other/openpathinscitechmathenginprimsecschtrans.pdf.

Uhden, O., Karam, R., Pietrocola, M., & Pospiech, G. (2011). Modelling mathematical reasoning in physics education. *Science & Eduaction, 21*(4), 485-506.

Vallerand, R. J., & Reid, G. (1984). On the causal effects of perceived competency on intrinsic motivation: A test of cognitive evaluation theory. *Journal of Sport Psychology, 6*, 94–102.

Vansteenkiste, M., & Ryan, R. M. (2013). On psychological growth and vulnerability: Basic psychological need satisfaction and need frustration as a unifying principle. *Journal of Psychotherapy Integration, 23*, 263–280.

Vansteenkiste, M., Simons, J., Lens, W., Sheldon, K. M., & Deci, E. L. (2004). Motivating learning, performance, and persistence: The synergistic role of intrinsic goals and autonomy support. *Journal of Personality and Social Psychology, 87*, 246–260.

Vansteenkiste, M., Williams, G. C., & Resnicow, K. (2012). Toward systematic integration between self-determination theory and motivational interviewing as examples of top-down and bottom-up intervention development: Autonomy or volition as a fundamental theoretical principle. *International Journal of Behavioral Nutrition and Physical Activity, 9*, 23. doi:10.1186/1479-5868-9-23.

Von Glaserfeld, E. (1993). Questions and answers about radical constructivism. In K. Tobin (Ed.), *The Practice of Constructivism in Science Education*. Hillsdale, NJ: Lawrence Erlbaum.

Von Schomberg, R. (2011). Prospects for technology assessment in a framework of responsible research and innovation. In M. Dusseldorp, & R. Beecroft (Eds.), *Technikfolgen abschätzen lehren. Bildungspotenziale transdisziplinärer Methoden* (pp. 39–61). Wiesbaden: Vs Verlag.

Vygotsky, Lev S. (1978). *Mind in Society: The Development of Higher Mental Processes*. Cambridge, MA: Harvard University Press.

Weinstein, N., Przybylski, A. & Ryan, R. (2012). The index of autonomous functioning: Development of a scale of human autonomy. *Journal of Research in Personality, 46*(4), 397–413.

Wenno, I. H. (2015). The correlation study of interest at physics and knowledge of mathematics basic concepts towards the ability to solve physics problems of 7th grade students at junior high school in Ambon Maluku Province, Indonesia. *Education Research International, 2015*, Article ID 396750, 6 pages, 2015. doi:10.1155/2015/396750.

Whitehead, A. H. (1929). *The Aims of Education*. Cambridge: Cambridge University Press.

Wilson, B. G., & Myers, K. M. (2000). Situated cognition in theoretical and practical context. In D. H. Jonassen & S. M. Land (Eds.), *Theoretical Foundations of Learning Environments* (pp. 57–88). Mahwah, NJ: Erlbaum.

Winn, F. S., Sarason, R. Fruland, P. O., & Lee, Y. L., (2006). Learning oceanography from a computer simulation compared with direct experience at sea. *Journal of Research in Science Teaching, 43*(1), 25–42.

Wolf, S. J., & Fraser, B. J. (2008). Learning environment, attitudes and achievement among middle-school science students using inquiry-based laboratory activities. *Research in Science Education, 38*, 21–341.

Wong, E. H., Wiest, D. J., & Cusick, L. B. (2002). Perceptions of autonomy support, parent attachment, competency and self-worth as predictors of motivational orientation and academic achievement: An examination of sixth-and-ninth-grade regular education students. *Adolescence*, 37(2002), 255–266.

Wood, J. T. (2013). *Interpersonal Communication. Everyday Encounters.* Canada: Wadsworth.

Zacharia, Z., Olympiou, G., & Papaevripidou, M. (2008). Effects of experimenting with physical and virtual manipulatives on students' conceptual understanding in heat and temperature. *Journal of Research in Science Teaching, 45*(9), 1021–1035.

Zacharia, Z., & Olympiou, G. (2010). Physical versus virtual manipulative experimentation in physics learning. *Learning and Instruction, 21*(3), 317–331.

Zeidler, D. L., & Nichols, B. H. (2009). Socioscientific issues: Theory and practice. *Journal of Elementary Science Education, 21*(2), 49–58.

Zusho, A., Pintrich, P. R., & Coppola, B. (2003). Skill and will: The role of motivation and cognition in the learning of college chemistry. *International Journal of Science Education, 25*(9), 1081–1094.

BALTISCHE STUDIEN ZUR ERZIEHUNGS- UND SOZIALWISSENSCHAFT

Herausgegeben von Gerd-Bodo von Carlsburg, Algirdas Gaižutis und Airi Liimets

Band 1 Gerd-Bodo Reinert / Irena Musteikienė (Hrsg.): Litauische Gespräche zur Pädagogik. Humanismus – Demokratie – Erziehung. 1999.

Band 2 Gerd-Bodo Reinert / Irena Musteikienė (Hrsg.): Litauische Gespräche zur Pädagogik II. Staat und Schule. 2000.

Band 3 Jaan Mikk: Textbook: Research and Writing. 2000.

Band 4 Gerd-Bodo Reinert / Irena Musteikienė (Hrsg.): Erziehungswandel und moderne pädagogische Verfahren. 2001.

Band 5 Airi Liimets (Hrsg.): Integration als Problem in der Erziehungswissenschaft. 2001.

Band 6 Gerd-Bodo Reinert / Irena Musteikienė (Hrsg.): Wissenschaft – Studium – Schule auf neuen Wegen. 2002.

Band 7 Palmira Jucevičienė / Gediminas Merkys / Gerd-Bodo Reinert: Towards the Learning Society. Educational Issues. 2002.

Band 8 Gerd-Bodo Reinert / Irena Musteikienė (Hrsg.): Bildung im Zeitalter der Informationsgesellschaft. 2003.

Band 9 Dalia Ohlinger: Argumentation in der Erst- und Fremdsprache. Pragmalinguistische und grammatikalische Aspekte anhand von Argumentationen deutscher und litauischer Studierender. 2003.

Band 10 Inga Šukevičiūtė: Interkulturelle Interferenzen im Bereich Wirtschaftskommunikation. Zur Analyse von Geschäftsbriefen im Vergleich Deutschland, Litauen und Russland. 2004.

Band 11 Gerd-Bodo Reinert von Carlsburg / Irena Musteikienė (Hrsg./eds.): Innovation durch Bildung. Innovation by Education. 2004.

Band 12 Gerd-Bodo von Carlsburg / Palmira Jucevičienė / Gediminas Merkys (eds.): Learning and Development for Innovation, Networking and Cohesion. 2005.

Band 13 Gerd-Bodo von Carlsburg / Irena Musteikienė (Hrsg./eds.): Bildungsreform als Lebensreform. Educational Systems Development as Development of Human Being. 2005.

Band 14 Gerd-Bodo von Carlsburg (Hrsg./ed.): Entwicklung erziehungswissenschaftlicher Paradigmen: Theorie und Praxis. Development of Educational Paradigms: Theory and Practice. 2007.

Band 15 Gerd-Bodo von Carlsburg (Hrsg./ed.): Bildungs- und Kulturmanagement. The Management of Education and Culture. 2008.

Band 16 Jaan Mikk / Marika Veisson / Piret Luik (eds.): Reforms and Innovations in Estonian Education. 2008.

Band 17 Gerd-Bodo von Carlsburg (Hrsg./ed.): Qualität von Bildung und Kultur. Theorie und Praxis. The Quality of Education and Culture. Theoretical and Practical Dimensions. 2009.

Band 18 Denkkulturen. Selbstwerdung des Menschen. Erziehungskulturen. Festschrift für Professor Dr. Dr. h.c. Dr. h.c. Heino Liimets. Herausgegeben von Airi Liimets. 2010.

Band 19 Ellu Saar (ed.): Towards a Normal Stratification Order. Actual and Perceived Social Stratification in Post-Socialist Estonia. 2011.

Band 20 Marika Veisson / Eeva Hujala / Peter K. Smith/ Manjula Waniganayake / Eve Kikas (eds.): Global Perspectives in Early Childhood Education. Diversity, Challenges and Possibilities. 2011.

Band 21 Airi Liimets / Marit Mäesalu (eds.): Music Inside and Outside the School. 2011.

Band 22 Gerd-Bodo von Carlsburg (Hrsg./ed.): Enkulturation durch sozialen Kompetenzerwerb. Enculturation by Acquiring of Social Competences. 2011.

Band 23 Reet Liimets: Ich als raumzeitliches Konstrukt. Die Fiktionen vom Leben der estnischen und deutschen Jugendlichen. 2012.

Band 24 Raivo Vetik (ed.): Nation-Building in the Context of Post-Communist Transformation and Globalization. The Case of Estonia. 2012.

Band 25 Airi-Alina Allaste (ed.): ‚Back in the West'. Changing Lifestyles in Transforming Societies. 2013.

Band 26 Gerd-Bodo von Carlsburg (Hrsg./ed.): Bildungswissenschaft auf der Suche nach globaler Identität. Educational Sciences in Search of Global Identity. 2013.

Band 27 Airi Liimets / Marika Veisson (eds): Teachers and Youth in Educational Reality. 2014.

Band 28 Gerd-Bodo von Carlsburg / Thomas Vogel (Hrsg./eds.): Bildungswissenschaften und akademisches Selbstverständnis in einer globalisierten Welt. Education and Academic Self-Concept in the Globalized World. 2014.

Band 29 Marika Veisson / Airi Liimets / Pertti Kansanen / Edgar Krull (eds.): Tradition and Innovation in Education. 2015.

Band 30 Robertas Jucevičius / Jurgita Bruneckienė / Gerd-Bodo von Carlsburg (eds.): International Practices of Smart Development. 2015.

Band 31 Gerd-Bodo von Carlsburg (Hrsg./ed.): Strategien der Lehrerbildung / Zur Steigerung von Lehrkompetenzen und Unterrichtsqualität. Strategies for Teacher Training / Concepts for Improving Skills and Quality of Teaching. 2016.

Band 32 Gerd-Bodo von Carlsburg (Hrsg./ed.): Denk- und Lernkulturen im wissenschaftlichen Diskurs / Cultures of Thinking and Learning in the Scientific Discourse. 2017.

Band 33 Palmira Pečiuliauskienė / Valdemaras Aleksa: Motivation of New Generation Students for Learning Physics and Mathematics. 2018.

www.peterlang.com

www.ingramcontent.com/pod-product-compliance
Ingram Content Group UK Ltd.
Pitfield, Milton Keynes, MK11 3LW, UK
UKHW041826031026
15587UKWH00002B/601

9 783631 746110